T0290887

Practical Realist
Philosophy of Science

Practical Realist Philosophy of Science

Reflecting on Rein Vihalemm's Ideas

Edited by Ave Mets, Endla Lõhkivi, Peeter Müürsepp, and Jaana Eigi-Watkin

LEXINGTON BOOKS

Lanham • Boulder • New York • London

Published by Lexington Books
An imprint of The Rowman & Littlefield Publishing Group, Inc.
4501 Forbes Boulevard, Suite 200, Lanham, Maryland 20706
www.rowman.com

86-90 Paul Street, London EC2A 4NE

British Library Cataloguing in Publication Information Available

Library of Congress Cataloging-in-Publication Data

Names: Mets, Ave, editor. | Lõhkivi, Endla, 1962- editor. | Müürsepp, Peeter, 1961-
    editor. | Eigi-Watkin, Jaana, editor.
Title: Practical realist philosophy of science : reflecting on Rein Vihalemm's ideas /
    edited by Ave Mets, Endla Lõhkivi, Peeter Müürsepp, and Jaana Eigi-Watkin.
Description: Lanham : Lexington Books, [2024] | Includes bibliographical references
    and index.
Identifiers: LCCN 2024010658 (print) | LCCN 2024010659 (ebook) |
    ISBN 9781666937220 (cloth) | ISBN 9781666937237 (ebook)
Subjects: LCSH: Realism. | Science—Philosophy. | Vikhalemm, R. A.—Influence.
Classification: LCC Q175.32.R42 P73 2024 (print) | LCC Q175.32.R42 (ebook) |
    DDC 501—dc23/eng/20240324
LC record available at https://lccn.loc.gov/2024010658
LC ebook record available at https://lccn.loc.gov/2024010659

Contents

Introduction

Ave Mets, Endla Lõhkivi, Peeter Müürsepp, Jaana Eigi-Watkin

Rein Vihalemm (1938–2015) was an Estonian philosopher of science and philosopher of chemistry, who in his later years, in the first decade of the twenty-first century, developed a unified framework of practical realism—a new kind of realist approach in the philosophy of science.[1,2] Here we provide an overview of some of Vihalemm's intellectual influences along his long road toward becoming an original thinker in the philosophy of science and one of the founders of the new philosophical discipline, philosophy of chemistry. We mainly look into the intellectual landscape of the Soviet Union, within which Vihalemm evolved as a philosopher, and the Hegelian and Marxist influences he employed to conceptualize science. The aim is to show just one small part of the complex influences he experienced as a philosopher of his time and place. Additional details about his philosophical influences from Kant, Marx, and Heidegger can be gleaned from his own statements published in *Acta Baltica Historiae et Philosophiae Scientiarum* (Vihalemm 2013). The Western analytic and pragmatic philosophies, which strongly influenced his thinking as well, will be introduced more cursorily. Further information about these influences can be found in his own English-language publications[3] and a handful of already published characterizations by his colleagues (see Müürsepp et al. 2019; Lõhkivi and Mets 2019).

Having obtained his first academic degree in chemistry, Vihalemm remained influenced by his primary discipline throughout his career as a philosopher and historian of science. Chemistry inspired him to introduce the term φ-science in scholarly discourse, a model of science proper designed after Galilean physics. The "φ" stems from "physics," exact science with its specific methodology which Vihalemm called constructive-hypothetico-deductive. He added "constructive" to the already established term "hypothetico-deductive," used to describe such sciences, based on the following considerations. First, these sciences formulate theories not directly about nature, but about models constructed to stand for nature, in a way, and directly defined by mathematics that is at the core of these sciences. Second, this construction is carried out by creating laboratory situations, which themselves are designed (pre-calculated) based on mathematical schemata and predictions. (One could also add that mathematics, the language of exact sciences, is constructed too.) Ronald Giere's constructivist and model-based understanding of science significantly influenced Vihalemm in formulating this aspect of science. The model of science, which Vihalemm called φ-science, is simultaneously an attempt to tackle the demarcation problem and an explanation to why physics is revered as the epitome of science. For him, the notion of laws of nature was also related to mathematical sciences—namely, it is the mathematically formulated laws of physics and physics-like sciences. Chemistry, according to this conceptualization, does have some important features of φ-science, but it cannot do without the other, non-φ-scientific part of it, which is the complex world of chemical substances and their properties, relevant in chemistry, that cannot be pre-calculated and can only be studied in material experiments. Vihalemm called this method of non-φ-sciences, stemming from natural history—classifying-historico-descriptive.[4] Thus, chemistry, according to Vihalemm, has a dual character.[5]

This influence of chemistry can be felt in the core ideas of practical realism as well. Vihalemm (2015, 100) saw it as an alternative to both antirealism and standard scientific realism. He presented the core of practical realism in the form of five tenets:

1. Science does not represent the world "as it really is" from a god's-eye point of view. Naïve realism and metaphysical realism have assumed the god's-eye point of view, or the possibility of one-to-one representation of reality, as an ideal to be pursued in scientific theories, or even as a true picture in the sciences.

2. The fact that the world is not accessible independently of scientific theories—or, to be more precise, paradigms (practices)—does not mean that Putnam's internal realism or "radical" social constructivism is acceptable.

3. Theoretical activity is only one aspect of science; scientific research is a practical activity and its main form is the scientific experiment that takes place in the real world, being a purposeful and critical theory-guided constructive, as well as manipulative, material interference with nature.

4. Science as practice is also a social-historical activity which means, amongst other things, that scientific practice includes a normative aspect, too. That means, in turn, that the world, as it is accessible to science, is not free of norms either.

5. Though neither naïve nor metaphysical, it is certainly realism, as it claims that what is "given" in the form of scientific practice is an aspect of the real

world. Or, perhaps more precisely, science as practice is a way in which we are engaged with the world. (Lõhkivi and Vihalemm 2012, 3)

For comparison, let us take a look at Rein Vihalemm's understanding of standard scientific realism, as he called it. It is "the conception according to which:

1. there is a mind-independent world (reality) of observable and unobservable objects (the metaphysical-ontological aspect),
2. the central notion is truth as correspondence between scientific statements (theories) and reality (the semantic aspect),
3. it is possible to obtain knowledge about the mind-independent reality (the epistemological aspect),
4. truth is an essential aim of scientific inquiry (the methodological aspect)." (Vihalemm 2015, 100)

According to Vihalemm, being an antirealist means rejecting at least one of these points. Therefore, a practical realist, as a realist, cannot reject any of these four points but can reinterpret them. Vihalemm claimed that practical realism evades the shortcomings of both antirealism and standard scientific realism. He would modify the first, metaphysical aspect of standard scientific realism as he understood it—namely, the mind-independent world, although extant, does not concern us or science for that matter. We only deal with the world as it manifests through our practices. Moreover, he rejected the import of metaphysics for understanding science; instead, scientific *practice* must be looked into to gain understanding about what science is. The second aspect is modified by suggesting that models act as intermediaries between scientific statements and the world, that is, scientific statements are not directly about the world but about models. Regarding both the second and the fourth aspects: truth is deflationary for Vihalemm, meaning that determining truth is equivalent to determining (via practices) how things are (Vihalemm 2011, 55). There seems to be no special problem with the epistemological aspect of standard scientific realism unless we accept a type of realism that assumes the God's-eye view—an impossibility, according to him.

To make proper sense of the background and essence of Vihalemm's conceptions of practical realism and a step leading toward it—the conception of φ-science, introduced above—we need to address the topics and thinkers that are not frequently in the forefront among philosophers of science today. Vihalemm matured as a philosopher in the Marxism-dominated Soviet system and had the opportunity to transition into the contemporary Western tradition only midway through his career. For this reason, he had to study philosophers, such as Hegel, whose works are generally quite unknown to

Western philosophers of science. Vihalemm's interest in Kant and Heidegger, not to mention Marx, is strongly related to his background and the sociohistorical circumstances he had to navigate through during his active years as a philosopher and historian of science.

In the Soviet Union, Marxism was officially considered the only correct view of the world, even though one could argue that the official ideological doctrine in the Soviet Union was not really a version of pure Marxism. Everyone involved in academic philosophy at least had to pretend to be Marxist. In informal communication with each other, Estonian philosophers of that period used the concept of "foreword Marxism," as merely citing the classics of Marxism-Leninism in the foreword to a philosophical treatise was often enough. Typically, the censors did not bother reading any further.

The philosophers of science managed to turn the situation in their favor, focusing on Marx's approach to practice and applying it to the analysis of the methodology of scientific research (see Mets 2019, 75n2). The main source was Marx's short philosophical piece called "Theses on Feuerbach." Normally, the first thesis has been pointed out as the core of the issue, the ideas that lead to the practical approach to science:

> The chief defect of all hitherto existing materialism . . . is that the thing, reality, sensuousness, is conceived only in the form of the object or of contemplation, but not as sensuous human activity, practice, not subjectively. Hence . . . the active side was developed abstractly by idealism—which, of course, does not know real, sensuous activity as such. (Marx 1845)

Here, Marx does not speak about science in particular. His approach to practice is a general epistemological one. Marx was interested in the interaction between objective and subjective reality. It is not about practice being the criterion of truth that the simplified interpretation suggests. Marx emphasizes the focal role of practice in making human cognition work and yield adequate information about the world out there. Just as in every other sphere of activity, this takes place in science.

In Marx's second thesis on Feuerbach, the idea of practice as the criterion of truth is introduced, stating that the objectivity of human thinking is a practical question. This implies that human practice is the means to validate truth, and thinking abstracted from practice and materiality is mere scholasticism (Marx 1845). This may possibly be the origin of Vihalemm's naturalist approach to science, along with the rejection of metaphysics. However, Vihalemm did not adhere only to the Marxist position. The attitude toward objective truth is different. For Marx, achieving objective truth remains the main goal of human cognition. In practical realism, however, this issue of truth almost does not deserve any attention whatsoever.[6]

Rein Vihalemm's knowledge of Marxism did not guide him directly to Kant, at least in the context of practical realism. Somewhat curiously, Martin Heidegger was needed to serve as a mediator for Vihalemm to revisit both Marx and Kant from a fresh viewpoint (see Vihalemm 2013). For obvious political reasons, in the 1990s, when Rein Vihalemm started to consolidate his ideas into what he eventually came to call practical realism, Marxism was despised in the former Soviet space. On the other hand, the philosophy of Heidegger became very popular. Many philosophers on the territory of the former Soviet bloc were probably looking for an alternative to Marxism. As most of them had been trained in the Hegelian-Marxist rather than the analytic tradition, Heidegger was a natural choice. Heidegger's philosophy seemed to bring fresh possibilities that were accessible to thinkers educated in the earlier German tradition. Still, by today, Heidegger has fallen out of fashion. Interest in him seems to have started to decline already before the publication of the *Black Notebooks* and the repercussions that followed.

Let us note that Vihalemm, lacking fluency in German, interpreted Heidegger with the help of certain Anglophone thinkers rather than addressing the German existentialist directly. Among the original texts of Heidegger, Vihalemm was probably most influenced by "The Age of the World Picture," in which Heidegger positions modern science within his general outlook about the most basic areas of human life in the contemporary world. This term ("scientific picture of the world") existed in Vihalemm's idea of science already before, in a strikingly similar meaning. It is the world picture or ontology of a particular scientific discipline (here Vihalemm drew significant influences from Vyacheslav Stepin [see, e.g., Stepin 2005], one of the practice-oriented philosophers of science in the Soviet system, and from Thomas Kuhn's conception of paradigm), or a more general scientific ontology of the world. This is a practical, not metaphysical conception of ontology, drawn from and laying the grounds for scientific practices (experiments and observations).

The following observation is speculative, as there are no direct references in Vihalemm's writings to support it. However, it may well be that Heidegger's explanation of the role of mathematics in contemporary science (physics) left a mark: "Modern physics is called mathematical because, in a remarkable way, it makes use of quite specific mathematics. But it can proceed mathematically in this way only because, in a deeper sense, it is already itself mathematical" (Heidegger 1977, 118).

Compared to the Greek concept *ta mathēmata* (see Heidegger 1977, 118), mathematics has kept its position as the basis for understanding all things in modern science, albeit in a much more narrow and concrete sense—as the language of science. This change is part of the big shift from qualitative to quantitative thinking about nature in the early modern period. Thus, just like in Kant, there is the fundamental position of mathematics in Heidegger's understanding of modern science. In addition, the mathematical way of understanding reality is, in a sense, pre-given. The basis of human scientific cognition is mathematical for both Heidegger and Kant. However, we cannot say that it is a priori in the Kantian sense for Heidegger.[7] Both ideas influenced the development of Vihalemm's thought.

In Anglophone literature, Vihalemm found kindred interpretations of Heidegger in the works of Jeff Kochan and Tricia Glazebrook (e.g., Kochan 2011; Glazebrook 2001). With Kochan's help, Vihalemm claimed that Heidegger's *Dasein* [Being-in-the-World] can be interpreted as practice (see Vihalemm 2013). This interpretation might seem too loosely substantiated, but let us note that Kochan is interpreting Heidegger, keeping in mind the connection with Joseph Rouse's ideas. The latter is one of the recognized leaders of the practice-based approach to science and has strongly influenced Vihalemm as well, especially as a thinker who dismisses the role of metaphysics in understanding reality. This is how Rouse sees the core of the issue:

> The question is not how we get from a linguistic representation of the world to the world represented. We are already engaged with the world in practical activity, and the world simply is what we are involved with. The question of access to the world, to which the appeal to observation was a response, never arises. The important categories for characterizing the ways the world becomes manifest to us are therefore not the observable and unobservable. We must ask instead about what is available to be used, what we have to take account of in using it, and what we are aiming toward as a goal. (Rouse 1987, 143)

Rein Vihalemm's practical realism is a further development of his own conception of proper science that he called φ-science. The latter is strongly influenced by Kant's conception of proper science. In a way, it can even be called a contemporary specification of Kant's approach to the matter. However, there is an important philosophical difference between the two as far as the aims are concerned. Vihalemm did not aspire to an idealist view of the matter but rather a practical perspective. Vihalemm's (2013) article gives his detailed account on this; here, we provide some background information about his approach. Namely, in adapting Kant's idea of proper science, Vihalemm directly applies the Hegelian idea of "sublation" (*Aufhebung*). The term has a threefold meaning: "to cancel," "to preserve," and "to elevate," which is something very typical of Hegel's triadic dialectical thinking.

Let us note that knowledge of Hegelian dialectics was a required part of education of philosophers in the Soviet system for an obvious reason. Hegel's dialectics was considered vitally important as the basis of Marx's dialectical materialism, his theoretical worldview. The three laws of the dialectics—the struggle of the opposites, change of quantity into quality (qualitative leap), and negation of negation—were picked up by the Marxists as a general explanation not only of natural but of social development as well. It is important to bear in mind that there are three basic laws of the dialectics. For Hegel, the triad—thesis, antithesis, and synthesis—was the framework for understanding reality in the most general terms. Therefore, the meaning of the important (from Hegel's point of view) concept of *Aufheben* is clearly triadic as well. In the context of practical realism, this means that Kantianism should definitely not be neglected but brought further on a different level. It also has to be kept in mind that Hegel's dialectical triad is actually a development of Kant's thesis–antithesis approach in his theoretical philosophy. The model image of the dialectical development is the spiral wherein everything tends to repeat itself, though at a different level.

We thus conclude this brief review of the particular set of influences on Vihalemm, and the subsequent section will be devoted to an even briefer overview of some additional influences.

Another great force that intellectually shaped Vihalemm was the contemporary Western developments in the philosophy of science, starting with Thomas Kuhn as the most prominent representative in this respect. Kuhn's practical-historical approach and the concept of normal science and paradigms left perhaps the strongest impression on Vihalemm's philosophical ideas. The practicality of science and the historicity of practice (including scientific practice) appear among the five tenets of practical realism listed above. A sort of sedimentation of knowledge, both theoretical and practical, as the basis of exact and natural scientific research (the fixity of mathematical theories and classifications), corresponds in important respects to Kuhn's normal science and the paradigm. In the post-Soviet context, it was no easy task to convince the administrators of science and education about the relevance and importance of the history of science to the philosophy of science. Rather, the philosophy of science was either seen as an auxiliary to science or as an abstract, theoretical field similar to the philosophy of language. All the more remarkable is Vihalemm's achievement in introducing and inculcating this practical-historical style in Tartu. His edited volume, *Estonian Studies in the History and Philosophy of Science* in the Boston Studies in the Philosophy of Science series (Vihalemm 2001), serves as evidence of this.

Vihalemm's methodological distinguishing of sciences, whereby he emphasized that different kinds of science have their own specific needs and aims and, hence, are all equally valid and necessary (implying methodological pluralism of sciences), and also his practice-based approach raised both methodological and metaphysical issues and led to debates with colleagues.

Some of the more prominent among the debates are those with Hasok Chang and Olimpia Lombardi, who both shared Vihalemm's appreciation for scientific practice as the true grounds for comprehending science. The central issue of those discussions is pluralism of science, with its practical-political and metaphysical corollaries. Vihalemm regarded the paradigmatic or normal basis of science, mentioned above, as one of the pillars of science, and also crucial for the advancement of exact sciences. Thus, his view on pluralism did not extend as far as to suggesting ontological-practical pluralism within a single discipline (contrary to Chang's approach, see Vihalemm 2016), or its importance for metaphysics—the idealistic pluralist realism of entities (contrary to Lombardi, see, e.g., this volume).[8]

Vihalemm expanded his ideas on the demarcation problem, which he tackled by means of the concept of φ-science, to what he called post-non-classical science (following Stepin 2005). In this categorization, the nonclassical science would be quantum physics, highlighting explicitly the observer's active role in gaining and shaping observation results, including Heisenberg's uncertainty principles. This type of science, in principle, still retains some hopes for certainty and predictability of classical physics, even if it remains computationally not (yet) achievable. It also retains the time reversibility of classical physics, in that it allows calculating a system forwards and backwards in time, providing an excuse to deny the reality of time, or of the arrow of time, in (some) fundamental reality. The post-non-classical science, represented by chaos theory and highly regarded by Ilya Prigogine, in principle, gives up predictability and places the arrow of time on the foundation of the scientific understanding of the world (see, e.g., Näpinen 2002, 2015).

The broader relations between science and society were a relevant topic for Vihalemm for at least two considerations: science being a practical social activity, intertwined in general human development, and his own political (social-democratic) inclinations and activity. Concerning these broader relations, he advocated that social sciences support Karl Popper's idea of feedback-driven social engineering (which aligns with similar ideas expressed by Rein Taagepera [2008], with whom he debated). He leaned particularly strongly on Nicholas Maxwell's ideas of the social role of science and the necessity of wisdom in shaping this role (see Näpinen 2015).[9]

Many others in the Anglo-American traditions of philosophy of science inspired Vihalemm; two prominent examples are Alfred North Whitehead and Wilfrid Sellars. Among his contemporaries, some of the philosophers whom he saw as allies and had fruitful discussions with are, for instance, Kenneth Westphal, Ilkka Niiniluoto, Loren Graham, Rom Harré; already mentioned Ronald Giere, Sami Pihlström, Hasok Chang, Olimpia Lombardi; and other philosophers involved in constructivism and HPS traditions. Already during the Soviet era, when traveling outside of the Soviet Union was complicated,

Vihalemm visited Western universities and conferences, for instance all the International Congresses on Logic, Methodology, and Philosophy of Science (and Technology), or CLMPSTs, and networked with the philosophers there. He also participated in the instantiation of the International Society for the Philosophy of Chemistry in 1997.

In conclusion, we would like to point out that the best introduction to practical realism in Vihalemm's own words is his article "Towards a Practical Realist Philosophy of Science" (Vihalemm 2011). Vihalemm's discussion on Kantian influences can be found in his "Interpreting Kant's Conception of Proper Science in Practical Realism" (Vihalemm 2013).

THE VOLUME'S STRUCTURE AND CONTRIBUTIONS

The first part of the collection of articles, "(Back)grounds of the Practical Approach," examines the historical affinities and discrepancies with Vihalemm's ideas. Recognition of the importance of material-practical interactions with the world in conditioning and shaping human cognition can be traced back to a distant past. Applying this recognition to scientific cognition is understandably more recent, considering that the contemporary understanding of science is only a few centuries old. Scientific cognition, however, is not an entirely separate type of cognition but builds upon our everyday interactions and conceptualizations of the world. The first part of the book explores some strands of these backgrounds.

Juho Lindholm argues that John Dewey anticipated much of the post-Kuhnian philosophy of science, in which Rein Vihalemm was involved. Lindholm demonstrates how Vihalemm's five theses of practical realism can be found in Dewey, either implicitly or explicitly. Dewey understood science as concrete and practical problem-solving (rather than as a set of true propositions or theories) and scientists as engaging actants that partially determine this practice. He also understood knowledge as a "kind of action," which Lindholm interprets as a practice oriented to the technical regulation of the environment and society. Vihalemm did not provide a systematic defense for his five theses, but Lindholm maintains that it can be constructed on the basis of Dewey's philosophy. These claims have effects beyond the narrowly scientific context.

David Hommen offers a critical appraisal of Vihalemm's practical realism from the perspective of Ludwig Wittgenstein's conception of mind and meaning. Vihalemm does not explicitly refer to Wittgenstein, but his philosophy may have been influenced by Wittgenstein through the work of Kuhn. Hommen argues that Vihalemm's practical realism boils down to an antirealist position, since Vihalemm only acknowledges the reality of nature in its material sense, but not in the proper realist sense, wherein beings as determinate entities exist independently of human cognition. Wittgenstein's account of linguistic-cognitive practices is shown to provide the necessary resources to remedy this deficiency and construct a true practical realist philosophy of science.

Kenneth R. Westphal stresses the importance of focusing not only on detailed issues within the philosophy of science, but also on the larger context within which both scientific and philosophical inquiry is, and must be, pursued. He highlights how Vihalemm's practical realism about science is rooted in our broadly existential grounds as embodied, active, cognizant beings operating within our material and historical-social world. Practical realism thus counters much of the philosophy of science, as fostered by Quine, which focuses on formal, (meta)linguistic aspects of science and the language(s) of scientific theories with their reported theoretical constraints. Westphal buttresses Vihalemm's views with a fundamental, yet widely neglected feature of Carnap's conception of language: the use of "descriptive semantics" and "genuine object-sentences" in scientific inquiry, as these alone allow assessing issues of truth, accuracy, or (in)sufficiency of evidence by working scientists. These considerations strongly support the tenets of Vihalemm's practical realism.

The second part of the book, "Metaphysics of Scientific Practices," elaborates on these aspects of Vihalemm's conception of science as a practice that reveal certain discrepancies with his otherwise like-minded peers in metaphysical matters, such as ontology, natural kinds, and pluralism.

Olimpia Lombardi articulates the concept of ontology, especially pluralist ontology, which Vihalemm seemed to denounce and disagree with, concurrently with his rejection of the relevance of metaphysics to science. Lombardi shows why it makes sense to discern scientific ontology, which accompanies scientists' theories and practices and is thus uncontroversial as a methodological background, from a ground ontology (or ontics) which she calls categorical-conceptual frameworks. Thus, she rehabilitates a noumenal world that science can connect to via its practices, and thereby also a meaningful, nonspeculative metaphysics relevant for science.

Hasok Chang contributes a chapter that advances the development of practice-oriented realism in two ways. First, he argues for accounts of truth and reality that conceive of them as constituted through operationally coherent activities. These accounts are described as sharing the spirit of Vihalemm's practical realism and filling some lacunae in it. Second, Chang shows how the notions of epistemic activity and system of practice are useful for the historiography of science. To do so, he gives an extensive example of the four systems of practice he identified in the history of creating the first batteries and developing their theoretical accounts.

Bruno Mölder reconstructs Vihalemm's practical realism by exploring the realist positions of Hilary Putnam and Ilkka Niiniluoto. The analysis focuses on the differentiation of natural kinds and mental kinds, with respect to their dependence on interpretation. According to Mölder, interpretivism serves well for mental kinds, whereas for natural kinds, the dependence on human activities is seen as problematic. Mölder disagrees with Vihalemm's account of natural kinds as being contingent on models constructed within scientific practices and argues for individuative independence of natural kinds. This allows him to maintain the differentiation of natural and mental kinds.

The third part of the book, "Special Sciences," presents applications and critiques of Vihalemm's ideas in specific scientific fields, predominantly in the natural sciences, but also one in the humanities.

Sami Pihlström compares Vihalemm's practical realism to pragmatic realism in the context of (the philosophy of) the humanities, particularly in religious studies. Practice-based accounts of science allow for the serious consideration of the plurality of practices of inquiry. This is especially important in the study of the humanities and, as the case of religious studies demonstrates, both pragmatic realism and practical realism would help contextualize religious practices. The strength of pragmatic realism appears in its reflexivity, but as Pihlström concludes, practical realism and pragmatic realism belong to the same research program in the philosophy of science and humanities, and thus should be developed together.

Klaus Ruthenberg discusses why philosophy of chemistry remains mostly neglected by philosophy of science in general. The main thread in Ruthenberg's chapter is a critical evaluation of several "theses" or explanations for this neglect. Intertwined with this, there is a discussion of different self-understandings of chemistry. Ruthenberg argues that the historically earlier understanding of chemistry as the science of substances, their qualities and production, remains crucial for understanding its character and philosophical interest. Thus, Ruthenberg sees connections with Vihalemm's philosophy: chemistry needs to be understood primarily through its experimental practices and its non-φ side.

Alexander Pechenkin and Apostolos Gerontas apply the framework of the φ- and non-φ-science, and chemistry's dual nature according to this framework, using the example of Belousov–Zhabotinsky reaction, a nonlinear chemical oscillator. This oscillator originates in biology, a non-φ-science, as it models the Krebs cycle in aerobic organisms (here, chemistry is also a non-φ-science as it describes and classifies natural phenomena). Additionally, however, it found a φ-scientific account in thermodynamics, specifically from the mathematical description of dissipative structures by Ilya Prigogine. Thus, Pechenkin and Gerontas present an apt example to illustrate the dual nature of chemistry.

Ave Mets takes under scrutiny the predominant activity of chemistry—producing plurality of new substances, and in her analysis focuses on the "constructive" element of Vihalemm's characterization of φ-sciences as applied to this aspect of chemistry. She further considers Ronald Giere's perspective on the construction of models, complementing Vihalemm's ideas. She shows that neither Vihalemm's conception of φ- and non-φ-sciences nor Giere's model-based approach, which are both supposed to encompass various types of sciences, can properly embrace this specific aspect of chemistry.

Apostolos Gerontas investigates the changes in the discipline of biology over the past decades from the perspective of Vihalemm's model of φ-science. Biology has increasingly adopted concepts, techniques, and instruments from chemistry and biology and expanded its field of study to molecular scales, for example, molecular biology and genetics. This suggests that, like chemistry, biology may have gained a dual nature, in Vihalemm's terms, via the process of chemicalization and physicalization of its domains. Gerontas scrutinizes the historically pivotal moments of these processes.

NOTES

1. This project has been supported by the European Union through the European Regional Development Fund (Centre of Excellence in Estonian Studies) and Estonian Research Council grant PRG462. We are grateful to all the reviewers whose valuable recommendations have helped to enhance the quality of the contributions. Images were created by Margus Evert, index by Raul Veede, and language editing done by Kait Tamm. We thank The University of Chicago Press for the permission to recreate figure 10.1, John Wiley and Sons for the permission to recreate figure 10.2(a), the Geis Archives for the permission to recreate figure 10.3(b), journals *Studia Philosophica Estonica* and *Philosophia Scientiæ* for permissions to use substantial parts of Vihalemm (2012) and Vihalemm (2015), respectively, and Triin Vihalemm, Peeter Vihalemm, and Marju Lauristin for the permission to use Arno Vihalemm's painting for the cover image.

2. See Sutrop (2015) for Rein Vihalemm's role in the Estonian philosophy landscape.

3. See his list of publications on the website of the Estonian Research Information System at https://www.etis.ee/CV/Rein_Vihalemm/eng/.

4. See Näpinen (2015) for a special exposition of φ-science; Mets (this volume) for a detailed criticism of some aspects of this concept. Lamża (2010) develops the idea of chemistry as a non-φ-science, with an emphasis on the "historical" aspect of it (out of the "classifying-historico-descriptive" of its methodology), providing a detailed discussion of the evolutionary aspects of chemical elements.

5. For discussions of laws of nature in chemistry, influenced by or led with participation of Vihalemm, see, for example, Vihalemm (2003, 2005), Christie and Christie (2003), and Tobin (2012).

6. Niiniluoto, whom Vihalemm saw as an ally in conceptualizing science, compares in Niiniluoto (2019) his and Vihalemm's ideas of truth and its role in science: this is a significant disagreement between them, as Niiniluoto adheres to the idea that truth as correspondence between articulations and the world is important in science, while Vihalemm dismisses the role of truth in science.

7. Rein Vihalemm definitely read the "Question Concerning Technology" as well. It is one of the first works by Heidegger to be translated into Estonian (Heidegger 1989), and it influenced many local philosophers despite the somewhat peculiar language usage by the translator. There was the option to consult the Russian translation for help.

8. See Manafu (2012) for some support of Vihalemm's rejection of ontological pluralism based on scientific practices; Pihlström (2012) includes a brief remark about Vihalemm's indifference toward metaphysical matters, specifically Kantian transcendentalism.

9. See Müürsepp (2011) for a development of the idea of knowledge, based on Vihalemm's conceptions of φ-science and non-φ-science along the lines of Maxwell; Näpinen (2011) for contrast, also employing Maxwell, among others, of φ-science knowledge and personal knowledge.

REFERENCES

Christie, John R., and Maureen Christie. 2003. "Chemical Laws and Theories: A Response to Vihalemm." *Foundations of Chemistry* 5: 165–74. https://doi.org/10.1023/A:1023631726532.

Glazebrook, Trish. 2001. "Heidegger and Scientific Realism." *Continental Philosophy Review* 34: 361–401. https://doi.org/10.1023/A:1013148922905.

Heidegger, Martin. 1977. "The Age of the World Picture." In *The Question Concerning Technology and Other Essays*, translated by William Lovitt, 115–54. New York, London, Toronto, and Sydney: Harper Perennial.

Heidegger, Martin. 1989. "Küsimus tehnika järele" (The Question Concerning Technology), translated into Estonian by Ülo Matjus. *Akadeemia* 6: 1195–228.

Kochan, Jeff. 2011. "Getting Real with Rouse and Heidegger." *Perspectives on Science* 19 (1): 81–115. https://doi.org/10.1162/posc_a_00026.

Lamża, Łukasz. 2010. "How Much History Can Chemistry Take?" *HYLE—International Journal for Philosophy of Chemistry* 16 (2): 104–20.

Lõhkivi, Endla, and Ave Mets. 2019. "Foreword." *A Story of a Science: On the Evolution of Chemistry*. Special issue of *Acta Baltica Historiae et Philosophiae Scientiarum* 7 (2): 3–5. https://doi.org/10.11590/abhps.2019.2.00.

Lõhkivi, Endla, and Rein Vihalemm. 2012. "Guest Editorial: Philosophy of Science in Practice and Practical Realism." *Studia Philosophica Estonica* 5 (2): 1–6. https://doi.org/10.12697/spe.2012.5.2.01.

Manafu, Alexandru. 2013. "Internal Realism and the Problem of Ontological Autonomy: A Critical Note on Lombardi and Labarca." *Foundations of Chemistry* 15: 225–28. https://10.1007/s10698-012-9165-x.

Marx, Karl. 1845. "Theses on Feuerbach." Translated: by Cyril Smith 2002, based on work done jointly with Don Cuckson. *Marx-Engels Archives*. Accessed October 22, 2023, http://www.marxists.org/archive/marx/works/1845/theses/index.htm.

Mets, Ave. 2019. "A Philosophical Critique of the Distinction of Representational and Pragmatic Measurements on the Example of the Periodic System of Chemical Elements." *Foundations of Science* 24: 73–93. https://doi.org/10.1007/s10699-018 -9567-x.

Müürsepp, Peeter. 2011. "Knowledge in Science and Non-Science." *Baltic Journal of European Studies* 1 (1(9)): 61–73.

Müürsepp, Peeter, Zhumagul Bekenova, Gulzhikhan Nurysheva, and Galymzhan Usenov. 2019. "From the Dual Character of Chemistry to Practical Realism and Back Again: Philosophy of Science of Rein Vihalemm." *Problemos* 96: 107–20. https://doi.org/10.15388/Problemos.96.9.

Näpinen, Leo. 2002. "Ilya Prigogine's Program for the Remaking of Traditional Physics and the Resulting Conclusions for Understanding Social Problems." *Trames* 6 (56/51) (2): 115–40. https://doi.org/10.3176/tr.2002.2.01.

Näpinen, Leo. 2011. "On the Unfitness of the Exact Science for the Understanding of Nature." *Baltic Journal of European Studies* 1 (1(9)): 74–82.

Näpinen, Leo. 2015. "The Premises and Limits of Science: The Ideas of Rein Vihalemm." *Acta Baltica Historiae et Philosophiae Scientiarum* 3 (2): 108–14. https:/ doi.org/10.11590/abhps.2015.2.06.

Niiniluoto, Ilkka. 2019. "Queries of Pragmatic Realism." *Acta Philosophica Fennica* 95: 31–43.

Pihlström, Sami. 2012. "Toward Pragmatically Naturalized Transcendental Philosophy of Scientific Inquiry and Pragmatic Scientific Realism." *Studia Philosophica Estonica* 5 (2): 79–94. https://doi.org/10.12697/spe.2012.5.2.06.

Pihlström, Sami. 2014. "Pragmatic Realism." In *Realism, Science, and Pragmatism*, edited by Kenneth R. Westphal, 251–82. Routledge Studies in Contemporary Philosophy. New York: Taylor and Francis.

Rouse, Joseph. 1987. *Knowledge and Power: Toward a Political Philosophy of Science*. Ithaca, NY: Cornell University Press.

Stepin, Vyacheslav. 2005. *Theoretical Knowledge*. Dordrecht: Springer.

Sutrop, Margit. 2015. "What Is Estonian Philosophy?" *Studia Philosophica Estonica* 8 (2): 4–64. https://doi.org/10.12697/spe.2015.8.2.02.

Taagepera, Rein. 2008. *Making Social Sciences More Scientific: The Need for Predictive Models*. New York: Oxford University Press. https://doi.org/10.1093/acprof :oso/9780199534661.001.0001.

Tobin, Emma. 2013. "Chemical Laws, Idealization and Approximation." *Science and Education* 22: 1581–92. https://10.1007/s11191-012-9445-9.

Vihalemm, Rein, ed. 2001. *Estonian Studies in the History and Philosophy of Science*. Boston Studies in the Philosophy of Science. Dordrecht: Kluwer. https://doi .org/10.1007/978-94-010-0672-9.

Vihalemm, Rein. 2003. "Are Laws of Nature and Scientific Theories Peculiar in Chemistry? Scrutinizing Mendeleev's Discovery." *Foundations of Chemistry* 5: 7–22. https://doi.org/ 10.1023/a:1021980526951.

Vihalemm, Rein. 2005. "Chemistry and a Theoretical Model of Science: On the Occasion of a Recent Debate with the Christies." *Foundations of Chemistry* 7: 171–82. https://doi.org/10.1007/s10698-005-0959-y.

Vihalemm, Rein. 2011. "Towards a Practical Realist Philosophy of Science." *Baltic Journal of European Studies* 1 (1(9): 46–60.

Vihalemm, Rein. 2012. "Practical Realism: Against Standard Scientific Realism and Anti-Realism." *Studia Philosophica Estonica* 5 (2): 7–22. https://doi.org/10.12697 /spe.2012.5.2.02.

Vihalemm, Rein. 2013. "Interpreting Kant's Conception of *Proper Science* in Practical Realism." *Acta Baltica Historiae et Philosophiae Scientiarum* 1 (2): 5–14. https://doi.org/10.11590/abhps.2013.2.01.

Vihalemm, Rein. 2015. "Philosophy of Chemistry Against Standard Scientific Realism and Anti-Realism." *Philosophia Scientiae* 19 (1): 99–113. https://doi.org/10 .4000/philosophiascientiae.1055.

Vihalemm, Rein. 2016. "Chemistry and the Problem of Pluralism in Science: An Analysis Concerning Philosophical and Scientific Disagreements." *Foundations of Chemistry* 18: 91–102. https://doi.org/10.1007/s10698-015-9241-0.

Part I

(BACK)GROUNDS OF THE PRACTICAL APPROACH

Chapter 1

John Dewey as a Precursor of Rein Vihalemm

Juho Lindholm

In philosophy of science, a practical turn is beginning to appear.[1] It is partly a reaction to mainstream analytic philosophy of science, realist (e.g., Boyd 1983, 1984, 1989, 1990, 1999; Niiniluoto 1987, 1999, 2018; Kitcher 1993, 2001; Psillos 1999; Papineau 2010), instrumentalist (e.g., van Fraassen 1980, 1989; Stanford 2006), historical rationalist (e.g., Laudan 1977, 1981, 1984, 1990; Lakatos 1978a, 1978b), and social constructivist (e.g., Barnes 1974; Bloor 1976; Collins 1981; Barnes and Bloor 1982; Fuller 1992) alike. Mainstream analytic philosophy of science treats science as an abstract system of representations (e.g., propositions, theories, model-theoretic structures) and conceives its task to be the analysis of the logical structure and language of such representations. The justification of theoretical knowledge is thought to consist in *rational reconstruction*. The concept of *experience* is reduced to mere observation; action is ignored without justification. Such a philosophy of science tacitly passes over concrete scientific practices—as if these practices were merely accidental, irrelevant, or even harmful for understanding scientific knowledge; and as if theories and observations made sense independently of practice.

An increasing number of scholars have attempted to describe science as a practice or an ensemble of practices. They do not *deny* the significance of theory and observation; but they *emphasize the practical function* of both. The practical turn began in the 1970s and 1980s in the sociology of scientific knowledge, science and technology studies, and certain other schools, for example, Ihde (1979, 1990, 1991, 1998), Latour and Woolgar ([1979] 1986), Knorr-Cetina (1981), Cartwright (1983), Hacking ([1983] 2010), Pickering (1984, 1995), Latour (1987), Rouse (1987, 1996, 2002), Radder (1988), Traweek (1988), Gooding, Pinch, and Schaffer (1989), Gooding (1990), Rheinberger (1992), Vihalemm (2001, 2011, 2012, 2013, 2015),

Baird (2004), Chang (2004, 2012), Giere (2006), Lõhkivi and Vihalemm (2012), and Currie (2018). According to Rouse (1987, ch. 2), Kuhn ([1962] 1996) should be included. This practice-centric approach has been called by various names: for example, as "new empiricism," "philosophy of science in practice," and "practice-based philosophy of science." The Estonian philosopher Rein Vihalemm (1938–2015) entitled his project as "practical realism."

In this chapter, I will examine the late Vihalemm's philosophical program. He criticized mainstream analytic philosophy of science—both realist and antirealist—for being too theory-centric; and scientific realism also for being too remote from actual scientific practice (Vihalemm 2011, 2012, 2013, 2015). Most of mainstream analytic philosophy of science focuses on physics, a highly theoretical science; hence, he argues, the theory-centrism. But that does not justify the exclusion of practice or of other sciences. Theorizing is indeed an important, maybe even indispensable part of science, but not the only one. For example, in chemistry, significant amounts of hands-on experience and practical doing and making come in. Widening the scope of philosophy of science would be important for discovering the value of practices and thus enriching our understanding of sciences—including physics.

There have been praxis philosophies before, which, however, were more general in scope and thus not emphatically philosophies of science: Marxism, pragmatism, and certain phenomenological philosophies like the early Heidegger ([1927] 1977), the later Husserl ([1936] 1976a, [1939] 1976b), and Merleau-Ponty ([1942] 1967, [1945] 2002). The later Wittgenstein ([1953] 2009, 1969) comes strikingly close to pragmatism. Ryle ([1949] 1951, ch. 2) made the distinction between *know-how* and *knowing that* (propositional knowledge). Polanyi ([1958] 1962, [1966] 1983) coined the notion of *tacit knowledge*. Arguably, already the mechanistic philosophy of the seventeenth and eighteenth centuries, which banished purpose from nature, was practical in the sense that it allowed humans to impose their own purposes on nature and hence to improve their condition technologically. Notwithstanding the existence of these alternative traditions in philosophy, mainstream analytic philosophy fails to do justice to them.

Classical pragmatism is an especially interesting case in point because it emerged from an attempt to understand the triumph of experimental science philosophically. The classical pragmatists conceived science as a process (practice) rather than a product (a system of practice-independent representations). Charles S. Peirce (1839–1914), who laid down the fundamental ideas of pragmatism, had firsthand experience of scientific practice. He characterized science as practice at least once in 1902 (CP 1.232–35; EP 2, 129–31). To my knowledge, he never abandoned this position in his later writings.[2] His collaborator William James (1842–1910) seems to have been uninterested in the philosophy of science per se.[3] But John Dewey (1859–1952) made a lot more out of Peirce's philosophy in his later works.[4]

Dewey conceived science as tangible and practical problem-solving rather than an accumulation of results: science was for him a mode of dynamic, concrete action in which living scientists engage rather than a static, abstract, and practice-independent system. Hence, rather than true justified representation (cf. Plato 1952, 97d–98a; 1977, 201c–d), he argued that knowledge is a "kind of action" (Dewey 1929b, 167).[5] He also pointed out that this identification of knowledge and action has already been made in physical science but not in philosophy.[6] As a fallibilist,[7] he considered theories and ideas to be hypothetical; and as a pragmatist, he also considered them to depend on practice because their function is to guide experiment. Hence theories are instrumental rather than final.[8] Thus, the end of science is not the contemplation of eternal and immutable truths but the intelligent and technical regulation of the human environment and society (Dewey 1916b, 1920, [1925] 1929a, 1929b, 1938). Dewey ([1925] 1929a, 308–9) pointed out that if the result of inquiry—propositions, theories, model-theoretic structures, and so on—is made the default state of affairs, then it remains a mystery how science ever arrived at them (cf. Lindholm 2021, 11). It turns out that conceiving science as practice solves this problem. If science consisted merely of the formal and rational reconstruction of its results, scientific practices would become accidental and unintelligible. But Dewey solves this problem by inverting it: it is the results which become unintelligible if practices are removed from the picture (see sections "Dewey's Epistemology" and "A Defense of Pragmatist Epistemology").

Vihalemm does not cite Dewey, but many of his theses are almost as if written by the latter. The only differences seem to be that (1) in his scientific publications, Vihalemm remains a philosopher of science,[9] but Dewey has explicit moral and social overtones in, and purposes for, his epistemological project, making it apparently wider in scope; and (2) Vihalemm's definition of practical realism remains largely programmatic and does not explain in detail why it would be better than mainstream analytic philosophy of science, while Dewey does provide such an argument (see sections "Dewey's Epistemology" and "A Defense of Pragmatist Epistemology").

In this chapter, I will argue that:

1. Dewey anticipated Vihalemm's practical realism in his middle and later philosophy, and
2. Dewey's (and Peirce's) theory of meaning provides an argument to convince mainstream analytic philosophers of science of the soundness of Vihalemm's practical realism.

I will first review Peirce's *pragmatic maxim,* which underlies Dewey's later works. Then I will present Vihalemm's theses in their original context:

as an alternative to mainstream analytic philosophy of science. I will review his theses first because, unlike Dewey, he formulated them clearly—although neither cites their opponents at all, thus leaving it unclear whom and what they oppose.[10] Dewey's exact position is sometimes difficult to extract from his prose, even though, at first glance, it seems easy to read. Dewey's theses and arguments are often vague, and therefore the way I present them here is only my interpretation. Alternative readings are possible. Anyway, Dewey can be understood more easily after an overview of Vihalemm's problematic. After the synopsis of Vihalemm, I will search for counterparts of each of his theses in Dewey's works. After pointing out similarities, I will discuss their justification and what it implies regarding their opponents. Finally, I will make some conclusions.

THE PRAGMATIC MAXIM

The theoretical background which underlies Dewey's later philosophy is Peirce's *pragmatic maxim*. Peirce published its original formulation in his 1878 "How to Make Our Ideas Clear":

> Consider what effects, which might conceivably have practical bearings, we conceive the object of our conception to have. Then, our conception of these effects is the whole of our conception of the object. (CP 5.402; EP 1, 132)

Peirce elaborates:

> What a thing means is simply what habits it involves. . . . Thus, we come down to what is tangible and practical, as the root of every real distinction of thought, no matter how subtle [*sic*] it may be; and there is no distinction of meaning so fine as to consist in anything but a possible difference in practice. (CP 5.400; EP 1, 131)

Peirce's original formulation seems to have problems with *conditionals* and *counterfactuals*. Thus, he later added that he means *potential* practice: in order for a sign to have meaning, it suffices that it be *potentially* interpret*able* in practice (CP 2.92, 2.275, 5.18, 5.196, 5.425–27, 5.438, 5.453, 5.457; EP 2, 134–35, 145, 234–35, 340–41, 346, 354, 356). Short (2007, 173) calls this "the subjunctive version of pragmatism": the meaning of a proposition is how it *would* influence conduct—that is, our habits of action—*were* it believed and *had* we some practical purpose to which it was germane.

In a word, Peirce defines the *meaning* of a thing as the *potential practical effects* of that thing, and *potential practical effects* as *habits*. One can readily see from this formulation that meaning is not restricted to language (cf. EP 2, 221); in principle, *anything* that has potential practical effects can be meaningful. A concept is meaningless if nothing practical follows from it. A conceptual difference must make a practical difference. The discoveries of second-generation cognitive science support the pragmatist theory of meaning (Lakoff and Johnson 1999). The determination of the meaning of a thing is an *experimental* problem (cf. CP 5.465; EP 2, 400–401): one can empirically determine what a thing means by experimenting on how different organisms respond to it. I have explained Dewey's account on how (linguistic or nonlinguistic) meaning emerges from experiment in Lindholm (2021, 7; 2022, 694; 2023a, 16; 2023b; forthcoming).

At any rate, the purpose of the maxim is to dispel all senseless metaphysics[11] which theoretical concepts and distinctions may yield, especially in rationalist philosophy like Descartes's and Hegel's. It also can be understood as a restatement of Kant's ([1781/87] 1956) claim that concepts only apply to experience; now "experience" is generalized to encompass also action. Dewey (1916a, 163–78; 1916b, 136n1, 270–78, 388; [1925] 1929a, 3a, 246–47, 279–80, 283, 314, 344–46; 1929b, 172–73, 234; [1934] 1980, 22, 53, 56, 132, 246, 251; 1938, chs. I–V; 1941, 183–84) was emphatic that experience is *organism–environment interaction*—a public, observable and causal process that cuts across the subject–object dichotomy (Lindholm 2023a, 13–14); and also a *bidirectional* process in which the organism and the environment influence each other (Lindholm 2023a, 6–7, 23–27). This notion seems to appear already in Peirce (CP 1.324, 1.336).

VIHALEMM'S PRACTICAL REALISM

Vihalemm (2015, 100) argues that practical realism can avoid the shortcomings of both standard scientific realism and antirealism. He maintains that knowledge cannot be understood as a representation of the world that is independent of practice, and that practice cannot be comprehended outside the framework of the real world. By "standard scientific realism," Vihalemm (2012, 10; 2015, 100) means the conception according to which:

1. There is a mind-independent world (reality) of observable and unobservable objects (the metaphysical–ontological aspect);
2. The central notion is truth as correspondence between scientific statements (theories) and reality (the semantic aspect);
3. It is possible to obtain knowledge about the mind-independent reality (the epistemological aspect); and
4. Truth is an essential aim of scientific inquiry (the methodological aspect).

Vihalemm (2012, 10–11; 2015, 100–101) criticizes this account for being too theory-centric (*logocentric*, using my idiom) and too remote from scientific practice. It also raises the problem of the God's-eye point of view. The various antirealisms like empiricism/instrumentalism and constructivism fall foul of the first criticism, too. On the other hand, some antirealisms[12] have the advantage that they have produced numerous case studies, which describe scientific practices in detail.

Here Vihalemm does not explain in sufficient detail why these are problems. I will show in sections "Dewey's Epistemology" and "A Defense of Pragmatist Epistemology" below that these indeed are problems and cut very deep; and how Dewey's conceptual apparatus solves these problems.

On the other hand, practical realism maintains that:

1. Science does not represent the world "as it really is," from a God's-eye position;
2. The fact that the world is not accessible independently of theories—or to be more precise, independently of paradigms (practices)[13] developed by scientists—does not mean that internal realism (Putnam 1981, ch. 3) or "radical" social constructivism is acceptable;
3. Theoretical activity is only one aspect of science;[14] scientific research is a practical activity whose main form is scientific experiment; the latter, in turn, takes place in the real world itself, being a purposeful, constructive, manipulative, and material interference with nature—interference, which is, in a crucial way, theory-guided;
4. Science as practice is also a sociohistorical activity: among other things, this means that scientific practice includes a normative aspect which, in turn, implies that the world actually accessible to science is not free of norms either; and
5. Though neither naïve nor metaphysical, it is certainly realism as it claims that what is "given" in the form of scientific practice is an aspect of the real world (Vihalemm 2015, 102).

Obviously, there is no one-to-one correspondence between the theses of standard scientific realism and practical realism. It can be seen from the last thesis that Vihalemm's did not intend to *refute* standard scientific realism per se but to *modify* and *qualify* it. They are targeted at realists and antirealists alike. Hence, it would have been informative to develop the implications of practical realism so that their exact relation to standard scientific realism (or antirealism) would become evident. But that exceeds the scope of this chapter. It requires another inquiry to examine, to what extent practical realism preserves scientific realism and to what extent it adopts elements from antirealism.

DEWEY'S EPISTEMOLOGY

It turns out that in my reading, all five theses of practical realism can be found already in Dewey's middle and later philosophy.

1. Vihalemm's denial of God's-eye point of view is roughly identical to Dewey's (1929b, 23, 196, 204, 211, 213, 245, 291) rejection of the "spectator theory of knowledge." Humanity is continuous with the rest of nature (Dewey 1910, ch. I; 1929b, 246), which implies that the human mind and knowledge are products of evolution. And the mind is no disinterested observer outside the world: it is manifest in concrete engagement within the world,[15] of which observation is part. Otherwise, we, biological organisms, would not survive. If we were immaterial Cartesian egos without bodily needs, the "spectator theory of knowledge" might make sense. But having a point of view does not preclude attaining partial truths: the criterion of knowledge, or a "kind of action,"[16] is *success or failure*, which can be "objectively" determined even from a restricted point of view (cf. Lindholm 2023a, 8, 31–32). Dewey (1929b, 204–5) also argues that if nothing else then at least Heisenberg's *principle of indeterminacy* undermines the "spectator theory of knowledge." If mere observation effects an alteration in what is observed, knowledge cannot but be about the consequences rather than the antecedents of inquiry. Otherwise, inquiry defeats its purpose. We understand a phenomenon insofar as we are able to initiate, sustain, and terminate it ourselves (Dewey [1925] 1929a, 428; cf. Lindholm 2023a, 32, 37; Marx MEW 3, 7, Engels MEW 21, 276–77, Kuusinen 1959, 98–99, 111, Hintikka 1969, 19–34, and Hacking [1983] 2010, 22–24). Dewey (1929b, 166–67, 189–91, 211) claims that the object of knowledge is constructed by acting in nature, using imagination and creativity, rearranging already existing things, not merely observing as if one was somehow "outside" nature. Dewey (1929b, 191) also claims that the purpose of inquiry is not the verification of an idea or a hypothesis but what can be done with its results—opportunities for action or "affordances" (Gibson 1979).

2. As for the claim that that the world is not accessible independently of paradigms, understood as practices,[17] Dewey (1922, 32) comes close: the world is accessed through the "refractive medium" of habits. The classical pragmatists use the term "habit" in a technical sense that cuts across the subject–object division. On the basis of prior actions and observations, habits create expectations which influence how certain patterns or gestalts and not others are observed in experimental data. People with different habits may observe different patterns or gestalts

in the same data. Dewey could, perhaps, be better understood if "habit" is replaced with "practice." Then there is no difference between Dewey and Vihalemm. Everything is indeed interpreted; but if Dewey is correct, interpretation is based on habits or practices, of which theorizing may be one. Habits or practices are neither a redefinition of representations (like sense data) that intrude between us and nature, nor a redefinition of "subjectivity"; on the contrary, they are part of nature (Dewey 1916a, 54–58; 1922, 14ff.).[18] That dispenses with skeptical problems. Kuusinen (1959, 92, 100) correctly observes that things manifest their properties when they change;[19] that suggests that experiment—the deliberate institution of changes—provides access to these properties. Moreover, Dewey and Vihalemm also might have added that data being interpreted through practices does not necessarily mean that that interpretation be false. On the contrary, habits or practices, being adaptations to objective constraints, anchor interpretations to relatively stable structures in the world (Määttänen 2009, 2015). Verbal descriptions may display large variations up to complete arbitrariness, but habits or practices cannot form in a vacuum (cf. Lindholm 2023a, 28–29). Each practice has a purpose, and the fact that there are objective constraints provides both the conditions and the means for agents to attain these purposes. Hence practices cannot be completely arbitrary, though the world often allows certain freedom for their evolution.

3. As it can be seen from the previous point, Dewey also anticipated Vihalemm's third claim that theorizing is only one aspect of science. For Dewey, science is art and art is practice; science is the intelligent component in any art which is itself an art ([1925] 1929a, ch. IX; cf. Lindholm 2021). Theorizing can, of course, be one of these practices (cf. Rouse 1996, 127). Dewey denies that a theory could be even articulated qua theory if abstracted from practice (1916a, 169).[20] He argues that the traditional distinction between theory and practice is irrelevant with regard to that between intelligent and unintelligent practice (1922, 69, 77; [1925] 1929a, 358). Dewey (1916a, 163–78, 237, 317–22; 1929b, 79–80, 84–85, 124, 199, 220, 240–42, 271, 295; 1938, ch. IV) denied the sharp distinction between science and ordinary problem-solving.[21] We all use the scientific method—the *experimental* method—in problem-solving. Both scientific inquiry and everyday problem-solving conform to Peirce's *belief–doubt model of inquiry* (cf. Lindholm 2023a, 25–27). Or, to say the same thing from the opposite point of view, there is no method exclusively peculiar to science. Thus, Dewey can be understood as equating experience and experiment. Note also that both Dewey and Vihalemm recognized that experiments are not random but are guided by ideas (or, more narrowly, theories), which manifests the rationality

inherent in experimentation (e.g., Dewey 1916a, 188–92; [1925] 1929a, 53, 76, 160, 260–61, 314–15; 1929b, 167, 226–27, 277, 302–3). Scientific practice has its own standard of rationality which needs no external justification save the accomplishment of results. This is a refutation of the doctrine that action and reason be distinct.[22] In fact, action is guided by the central nervous system, and motor actions and reactions develop the central nervous system; therefore, action cannot possibly be distinct from reasoning. It could be said that in *any* intelligent act of experimentation, an implicit, unarticulated theory is manifest.

4. Dewey (1938, chs. I–V) preceded Vihalemm also in holding that science is sociohistorical and normative. He holds that inquiry always takes place within a biological and a social matrix.[23] It is one possible mode of interaction between an organism and environment (i.e., experience); and other organisms and their reactions are part of the environment (cf. Lindholm 2023b). Hence, social norms necessarily regulate (but do not necessarily determine) science. However, the norms of inquiry can also stem from *technical norms* (von Wright 1963; cf. Määttänen 2009, 33).[24] Moreover, against the thesis of value-independence of science, science *ignores* values, purposes, and qualitative objects *methodically* and *instrumentally* to facilitate certain operations but does not really *eliminate* them from ontology. Alas, how could they disturb inquiry if they were not real? After inquiry, the scientist returns to the realm of ordinary objects without change in their value. An object as viewed by ordinary experience and by science is the same; it is just treated with different purposes (Dewey 1929b, 102–7, 128–29, 131, 219–22, 240–42, 271, 295).

5. Dewey was a "realist," but not in the metaphysical sense. Unlike in the influential Parmenidean tradition, he did not oppose reality to appearance:[25] "[t]he world as we experience it is a real world" (Dewey 1929b, 295).[26] According to the pragmatic maxim, the predicate "real" is only applicable to what appears in experience. That can, perhaps, be understood as a version of *phenomenalism without subjectivism:* the world consists of phenomena, but they need not be conceived as ideal, "subjective," "internal" entities opposed to real, "objective," "external" entities (cf. Lindholm 2021, 9n58; 2023a, 19n41). Traditional empiricists like Locke and Hume considered experience as purely "subjective" sense data opposed to the "objective" world. On the other hand, Dewey modeled experience after concrete, causal interaction between an organism and its environment. If experience is equated with experiment, it must be understood correspondingly as a causal, public, observable process within nature, which itself can be studied experimentally—hence the later Dewey's naturalism and the reflexivity of his notion of science: its

method applies to itself. Thus, he would not have considered experience as private or exclusively mental. The mind is involved, of course, but he did not conceive mind as an entity outside nature but as a function or attribute or mode of the organism–environment interaction.[27] Thus, he would have said that all events which manifest the mind are public, observable, and causal and hence "real," "objective," and "external." In Dewey's scenario, there is no ontological asymmetry between experience and nature.[28] Hence empiricism is not opposed to realism. On the contrary, the former includes the latter.

There are certain other claims in common to Dewey and Vihalemm. Vihalemm (2012, 10; 2015, 102) emphasizes that objects are not self-identifying. Dewey ([1925] 1929a, 308–9) says the exact same thing. According to him, realists erroneously make the result of inquiry the default state of affairs (cf. Lindholm 2021, 11). This obscures the *problem of objectification:* we do not know a priori what the object of inquiry will be when we have complete knowledge about it (cf. Lindholm 2023a, 20). Excluding the possibility of teleological explanation, the result cannot influence inquiry in progress.

A DEFENSE OF PRAGMATIST EPISTEMOLOGY

I have argued above that Vihalemm criticized mainstream philosophy of science for being too theory-centric (*logocentric*, using my idiom). But Vihalemm does not explain in detail why theory is insufficient for understanding science. It is not enough even to address observations as well. Dewey (1916a, 169; 1929b, 112–13, 128–29, 136–38; 1938, 66–70) tells us why: if abstracted from practice, both theories and observations become meaningless. But Dewey does not tell in sufficient detail why understanding requires practice. In order to justify that claim, one needs recourse to Peirce's semiotics.

Neither observations nor theories mean anything inherently. They are just so many more things. And things, as such, do not have a mysterious power to "refer" outside themselves. In order to do that, they must be *sign-vehicles*, which are things to which the power to refer is conferred by their being related to two other things in a certain way. This is where an excursion to Peirce's semiotics comes in aid.

A distinction must be made between *sign-relation* and *sign-vehicle*. Peirce equivocally called both "signs." Usually the context reveals which one he means.[29] The latter is one of the relata of the former. The other two relata are an *object*[30] (i.e., what the sign-vehicle refers to) and a potential *interpretant*[31] (how the sign-vehicle is interpreted) (CP 2.92, 2.242, 2.274; EP 2, 13, 272–73, 290). The sign-vehicle must be a physical medium (EP 2, 326).

Another distinction must be made between a *dynamical object* and an *immediate object*. The latter denotes an object as represented by a sign: how that object appears at a given stage of inquiry. The former denotes the real object independently of how the sign represents it: how it appears at the conclusion of the inquiry, defined as the point where everything about the object is known, and no further increase in knowledge is possible. The immediate object is a part of the dynamical object. Hence, the distinction is only relative. The immediate object indicates how the rest of the dynamical object can be discovered by *collateral experience* (CP 6.318, 6.338, 8.178–79, 8.183, 8.314; EP 2, 404–9, 429, 480, 493–98).

The *interpretant* is "what the sign in its significant function essentially determines in its interpreter" (EP 2, 409) or "the mental effect which the sign-vehicle has upon the interpreter" (EP 2, 429). There are different kinds of interpretants, and they constitute the meaning category.

Peirce allows the interpretant to be another sign-vehicle which has the same dynamical object. That involves another interpretant, and so on. Thus, there may arise a *progressus* of sign-vehicles interpreting previous sign-vehicles. This process is called *semiosis*. It is potentially infinite, but it tends to termination (Short 2007, 91–150, 158, 171–74; Vehkavaara 2007, 263–64, 273). During semiosis, we retain direct access to the dynamical object. By directly interacting with it, we form a habit inductively (CP 2.643; EP 1, 198–99). In this process, the immediate object develops. Semiosis terminates in a "quality of feeling," "exertion" (i.e., action), or habit (CP 4.536, 8.332; MS 318). The dynamical object is understood in terms of habit. A thing is known by subsuming it under a universal; and a habit is a universal as it determines actual and possible actions.[32] Habit, or the *final logical interpretant*, is required to associate a sign-vehicle with a dynamical object (cf. Määttänen 2009, 90–102, 126–29). Semiosis ultimately terminates when the immediate object coincides with the dynamical object, that is, the dynamical object is known in all its aspects; all habitual encounters with the dynamical object have been exhausted; and no further increase in knowledge is possible.

Now, theories and observations are sign-vehicles. Both can only be understood in terms of habit (CP 6.481; EP 2, 447–48). Abstracted from practice, both cease to function as sign-vehicles and become meaningless. Thus, Dewey (1916a, 169) is correct in denying the possibility of the articulation of a theory qua theory independently of practice. The basic error of traditional theories of knowledge resides in the isolation and fixation of some phase of the whole process of inquiry in resolving problematic situations (Dewey 1929b, 171–80, 188, 289–90). The conclusion of an inference is not the conclusion of an inquiry (Dewey [1910] 1933, 100–101); it merely states the possibility of the solution of a problematic situation which somebody still must actualize in practice.

Moreover, in abstraction from purposive practice, we could not determine which observations and theories are important (Dewey 1929b, 171–80, 188; 1938, 66–70; cf. Lindholm 2023a, 19–23). A purpose acts as a filter. It provides an ability to attend to observations and concepts that are relevant with regard to it and to ignore the rest. Otherwise, we would be buried under a mass of irrelevant cognition and could not cope with the situation. We would risk a nervous breakdown, trying to pay attention to every minutest detail in our environment. This is the case in science, too. We cannot measure everything. And even if we could, not all measurements are signs of anything else. We would be wasting resources in recording uninformative data.

For this reason, experimentation has its own rationality: it is directed by a purpose, prior ideas, and prior theories which determine what it makes sense to try next. The negligence of practice indicates that philosophers of science consider it as an unimportant addendum.[33] This implies that they think that *rational reconstruction* exhausts scientific rationality. This implies that practices are arational—or even irrational. But Dewey (1916a, 39, 54–58, 62, 263–64, 319, 323, 400) denied the distinction between reason[34] and experience. On the contrary, reasoning is part of experience. He claimed that (scientific) practices are neither unintelligent nor irrational, nor are theories devoid of practical basis: ideas and theories guide subsequent experimentation but are themselves acquired by prior experimentation. They are hypothetical *conditionals* which claim that if something *were* true, then something else *would* follow (Dewey 1929b, 86, 163–65, 168, 273).

Hence, I conclude that practice is the vehicle of cognition, and theories and observations have cognitive value only as its non-independent parts.

DISCUSSION

The root of the problems of contemporary epistemology and philosophy of science seems to be the narrow notion that experience be observation—or, even worse, stimulus (Quine 1995). That notion is inherited from the tradition which stems at least from Aristotle (1935), if not earlier.[35] It is still prevalent: see any article in the semi-authoritative anthology of analytic epistemology edited by Sosa, Kim, Fantl, and McGrath (2008): *not a single author* mentions the epistemic value of practice (cf. Lindholm 2023a). This notion implies the passivity of the subject of knowledge, or "spectator theory of knowledge"—a notion which is simply unacceptable in post-Darwinian philosophy. No living animal could survive if it were adopted in practice. Taking the theory of evolution seriously forces one to make certain revisions in philosophy—that the mind has evolved to coordinate purposeful action by observations; theorizing is ancillary. Contemplation of theories is, indeed, a possibility in a society which is materially developed enough; but it is still an accidental property of the mind. Hence it should not be made the paradigm case by which all knowledge be judged.

Literally, by the criteria of standard scientific realism, Vihalemm's program counts as antirealism. But his fifth thesis indicates that he calls for a more liberal definition of "realism." I think this makes sense: the criteria for standard scientific realism are too demanding. Especially semantic realism seems problematic to me. I will not present detailed arguments against it here, but I will make two remarks. I am inclined to call any position "realism" or "realism *simpliciter*" if it recognizes the ontological aspect of realism—with or without the God's-eye point of view. Accordingly, I am inclined to call positions which additionally include some or all of the three other aspects of realism variants of realism, for which some more expressive titles can be made up, such as "ontologico-semantic realism" or "ontologico-epistemic realism."

There is, however, an opportunity for the criticism of the requirement of mind- or theory-independence of realism *simpliciter*. If Dewey (1916b, 35–36) is correct, the object of knowledge is the *product* of inquiry (or a "construct" or an "artifact") and thus profoundly dependent on the activity of the mind or on previous theories and ideas. But that does not necessarily make it unreal. For example, my desk and my computer on it are constructs. The process of their construction is obviously mind- or theory-dependent; but once brought into existence, they are as mind-independent as the earth below my feet. Moreover, the raw materials of which they are composed are mind-independent too. If all thinking creatures were removed from the universe, these things would remain intact.

CONCLUSION

If Dewey's philosophy is interpreted in the way proposed, the five theses of Rein Vihalemm's practical realism can be found in it. For Vihalemm, these theses were programmatic, and he provided no systematic defense for them. On the other hand, Dewey's philosophy, with the support of Peirce's semiotics, provides an argument for them. Hence, I claim that in order to appeal to other philosophers, Vihalemm's theses should be backed up by my argument. It shows that neither of the mainstream constituents of science, theory and observation, makes sense in abstraction from practice. This is a serious challenge to mainstream epistemology and philosophy of science.

Most importantly, my argument shows that there are fruitful alternatives to logocentric (representationalist) philosophy of science. Dewey's (and Peirce's) thought captures the same position that makes standard scientific realism so appealing: that the distinction between observables and unobservables makes no sense in the absence of distinction in practice. At the same time, they dispense with the metaphysical baggage of standard scientific realism.

One possibility for further inquiry, which I already indicated, is whether Vihalemm's practical realism and Dewey's operationalism are coherent. As I have equated the terms "habit" and "practice," another natural sequel to this inquiry is a comparison between John Dewey and Joseph Rouse. In spirit, their ideas are strikingly similar. But Rouse never cites Dewey. The former may have been influenced by the latter through Richard Rorty. And I have already mentioned three additional possible directions for inquiry: to what extent does practical realism preserve the doctrines of standard scientific realism and to what extent does it adopt elements from antirealism; the criticism of semantic, epistemic, and methodological realism; and the criticism of the mind-independence requirement of ontological realism.

NOTES

1. I thank the EU Regional Fund, Dora Plus, Estonian Research Council, and Grant no. 462.
2. On the other hand, Peirce claimed that "pure science has nothing at all to do with action" (CP 1.635; EP 2, 33). I believe that this is incoherent. What could a practice that has nothing at all to do with action look like? (cf. Lindholm 2021, 5).
3. See James (1890, 1902, [1897] 1907, [1907] 1916, 1909, 1912).
4. See Dewey ([1925] 1929a, 1929b, 1938).
5. Dewey is not explicit about what this claim means. I understand him saying that actions come in kinds, and these kinds amount to knowledge. Kinds of action could be simply called "habits" or "practices" (Lindholm 2021, 7). Dewey (1929b, 204–5) also says that knowledge is a kind of *inter*action.
6. Science has become experimental at least since the Middle Ages, but mainstream analytic philosophy of science, both in Dewey's time and nowadays, still ignores practice. Although the situation has improved since the beginning of the practical turn, this makes Dewey's ideas still pertinent to a degree.
7. In short, *fallibilism* is the position that *anything (but not everything at once) can be doubted, if positive reasons to do so arise.* See Peirce (CP 1.7, 1.135, 3.432, 4.531, 5.265, 5.367, 5.374–87, 5.416–17, 5.577, 5.582–84, 6.595; EP 1, 28–29, 112, 114–23; EP 2, 26, 44, 47–48, 336–37) and Dewey (1916b, 1929b, 1938).
8. This, however, does not preclude them from being true. To my knowledge, he never says in his middle and later works that theories and ideas have no truth-value, even though in *Logic* (1938, esp. 7–9) he prefers the term "warranted assertibility." He also subscribes to the correspondence theory of truth, though in an "operational" sense, which presumably means the rejection of transcendent entities like noumena (Dewey 1941, 178–79). It seems to me that each classical pragmatist understood

truth differently. For discussion, see Peirce (CP 5.407, 5.430, 5.553; EP 1, 139; EP 2, 379–80, 432–33), James ([1907] 1916, 64, 80, 198, chs. VI–VII; 1909, v–xx, chs. III, V–IX, XII–XIII), and Dewey (1916b, 240–1, 324–25; 1920, 155–60). See also Rouse (1987, 7–8), Lakoff and Johnson (1999, 6, 94–95, 98–106), and Short (2007, 333). For background, see Aristotle (1933, 1011b25), Thomas Aquinas ([1256–59] 1918, pt. 1, q. 16, a. 2, arg. 2), Husserl ([1900] 1975; [1901] 1984a; [1901] 1984b; 2002; 2005), Russell ([1918] 2010), and Wittgenstein (1922).

9. In his private life, Vihalemm was socially and politically active, but he did not show that in his publications.

10. To identify their opponents, I would characterize them under the title of theory-, logic-, and language-centrism, or what I call *logocentrism*: the notion that it would be sufficient to analyze the language and logic of scientific theories in order to understand science. Rouse (1996; 2002, ch. 4) calls this tradition *representationalism*. It includes scientific realism, empiricism/instrumentalism, historical rationalism, and social constructivism. I consider such an approach as a continuation of traditional rationalism, where "Reason" (with capital "R") is substituted with "language" or "logic."

11. Here, I emphasize the word *senseless* in order to make room for metaphysics that *does* make sense because its concepts arise from practice. The classical pragmatists believed that such metaphysics is possible.

12. Like Latour and Woolgar ([1979] 1986). See also Baird (2004, 7).

13. Rouse (1987, ch. 2) has argued that Kuhn's ([1962] 1996) "paradigm" can be interpreted as practice.

14. Rouse (1996, 127) has argued that theorizing itself is one peculiar practice among others (cf. Dewey 1916a, 169; 1922, 69, 77; [1925] 1929a, 358).

15. Thus, Dewey ([1925] 1929a, 158–59; [1934] 1980, 263) emphasizes that "mind" should be understood as an attribute of action rather than a substance (cf. Hickman [1990] 1992, 10).

16. See note 5.

17. Access via practices may involve theories. Hence, this claim accounts for the notorious thesis of the theory-ladenness of observation but is enlarged into a thesis of the practice-ladenness of observation. Theory-ladenness was made famous by Quine (1951). But this idea can be traced back to Kant's first *Critique* ([1781/87] 1956) (if not earlier), Peirce's 1903 *Harvard Lectures on Pragmatism* (CP 1.314–16, 5.14–81, 5.88–212, 5.77n; EP 2, 133–241), and Pierre Duhem ([1906] 1954). Arguably, the notion that the mind is active in observation was known already to Platonists, neo-Platonists, Augustinians, and Franciscans (Pasnau 1997).

18. See also Rouse (1987, chs. 4 and 7; 1996, chs. 5–9; 2002, chs. 5–9).

19. I believe that Kuusinen's claim can be derived from Newton's laws of motion and the conservation laws of physics. When a thing communicates anything by a physical signal to its surroundings (including human observers), it necessarily changes; and the reception of this signal, in turn, necessarily changes the surroundings (cf. Lindholm 2023a, 24). This, in turn, suggests process ontology (cf., e.g., Whitehead [1929] 1978). Kuusinen's claim might also be derivable from the pragmatic maxim: what a thing *is* is what it *does*.

20. I will give a semiotic argument for this claim in section "A Defense of Pragmatist Epistemology."

21. Peirce (CP 5.438–52, 5.494; EP 2, 346–54, 419–20) made the same point.

22. Dewey (e.g., 1916a, 39, 54–58, 62, 263–64, 319, 323, 400) seizes upon this point.

23. Ave Mets (in private communication) astutely pointed out that science also takes place within a technical matrix. Dewey (1938, ch. III) subsumes it under the social matrix.

24. Hence, a practice can, in principle, have its own criteria of success or failure independently of society. If an individual or a subgroup has the purpose of producing a certain effect, society can, of course, approve or disapprove of that, but whether and how that effect is obtained can be independent of it. Some practical problems are thus purely technical. For example, if there is a society that forbids the bringing forth of a certain product, people can still do that, when and where the prohibition is not or cannot be enforced.

25. Parmenides distinguished between the way of opinion (δόξα) and the way of truth (ἐπιστήμη) (Diels and Kranz 1960, 28A, 28B; cf. Diogenes Laërtios 1905, 384–85).

26. See also Dewey (1916b, ch. IX) and Rouse (1987, 7–8). This might arguably involve a problem with the second thesis. If the world is accessible only through the "refractive medium" of practices (cf. Dewey 1922, 32), how can it be real? This question, again, is beyond the scope of this chapter: I confine this inquiry to discover similarities between Dewey and Vihalemm and thus exclude the detection of possible contradictions. As a sketch of a reply, I suggest that the "refraction" which practices induce is not completely arbitrary but is an adaptation to objective constraints. Hence, it can be misleading only in a certain limited range. See Määttänen (2009, 2015).

27. This makes Dewey and possibly other pragmatists precursors of the 4E cognition theories, which conceive mind as embodied, embedded, enacted, and extended. See, for example, Menary (2010) and Newen, De Bruin, and Gallagher (2018).

28. See esp. Dewey ([1925] 1929a).

29. Sometimes Peirce called the sign-vehicle a *representamen*.

30. Here I mean a *dynamical* object, which I will define in the following paragraph.

31. The sign-relation can function even if the interpretant is only potential. This makes Peirce's mature pragmatism "subjunctive," using Short's (2007, 173) idiom.

32. Aristotle (1962, 17a35–b5) defined *universal* as something that can be predicated of many things. Habits are different kinds of universals: they can be *enacted* in many *situations*. See Lindholm (2023a, 27–30).

33. Karl Popper ([1934] 1953, 107) has succinctly expressed the theory-driven picture of science: the theoretician accomplishes everything significant; experiment is incidental. See also Rouse (1987, 96–97).

34. Dewey preferred to substitute "intelligence" for "reason" in order to take distance to the connotations associated with the latter by traditional philosophical systems.

35. See also Pasnau (1997) for the medieval dispute about whether the mind is active in observation.

REFERENCES

Aristotle. 1933. *The Metaphysics, Books I–IX.* Greek–English edition. Translated by Hugh Tredennick. London and New York: William Heinemann Ltd. and G. P. Putnam's Sons.

Aristotle. 1935. On the Soul. In *On the Soul; Parva Naturalia; On Breath.* Greek–English edition. Translated by W. S. Hett, 8–203. London and Cambridge, MA: William Heinemann Ltd. and Harvard University Press.

Aristotle. 1962. On Interpretation. In *The Categories; On Interpretation; Prior Analytics.* Greek–English edition. Translated by Harold P. Cook, 114–80. London and Cambridge, MA: William Heinemann Ltd. and Harvard University Press.

Aquinas, Thomas. [1256–59] 1918. *Quaestiones disputatae de veritate, quaestio IX,* edited by A. Dyroff. Bonna: Petrus Hanstein.

Baird, Davis. 2004. *Thing Knowledge: A Philosophy of Scientific Instruments.* Berkeley, CA: University of California Press. https://doi.org/10.1525/9780520928206.

Barnes, Barry S. 1974. *Scientific Knowledge and Sociological Theory.* London and Boston, MA: Routledge and Kegan Paul.

Barnes, Barry, and David Bloor. 1982. "Relativism, Rationalism, and the Sociology of Knowledge." In *Rationality and Relativism,* edited by M. Hollis and S. Lukes, 21–47. Cambridge, MA: MIT Press.

Bloor, David. 1976. *Knowledge and Social Imagery.* London: Routledge.

Boyd, Richard N. 1983. "On the Current Status of the Issue of Scientific Realism." *Erkenntnis* 19 (1/3): 45–90. https://doi.org/10.1007/BF00174775.

Boyd, Richard N. 1984. "The Current Status of Scientific Realism." In *Scientific Realism,* edited by J. Leplin, 41–82. Berkeley, CA: University of California Press. https://doi.org/10.1525/9780520337442-004.

Boyd, Richard N. 1989. "What Realism Implies and What It Does Not." *Dialectica* 43 (1–2): 5–29. https://doi.org/10.1111/j.1746-8361.1989.tb00928.x.

Boyd, Richard N. 1990. "Realism, Approximate Truth and Philosophical Method." In *Scientific Theories,* Minnesota Studies in the Philosophy of Science, vol. 14, edited by C. W. Savage, 355–91. Minneapolis, MN: University of Minnesota Press.

Boyd, Richard N. 1999. "Kinds as the 'Workmanship of Men': Realism, Constructivism, and Natural Kinds." In *Rationalität, Realismus, Revision: Proceedings of the Third International Congress, Gesellschaft für Analytische Philosophie,* edited by J. Nida-Rümelin, 52–89. Berlin: de Gruyter. https://doi.org/10.1515/9783110805703.52.

Cartwright, Nancy. 1983. *How the Laws of Physics Lie.* Oxford: Oxford University Press. https://doi.org/10.1093/0198247044.001.0001.

Chang, Hasok. 2004. *Inventing Temperature: Measurement and Scientific Progress.* Oxford and New York: Oxford University Press. https://doi.org/10.1093/0195171276.001.0001.

Chang, Hasok. 2012. *Is Water H_2O? Evidence, Realism and Pluralism.* Boston Studies in the Philosophy of Science. Dordrecht, Heidelberg, New York, and London: Springer. https://doi.org/10.1007/978-94-007-3932-1.

Collins, Harry M. 1981. "Stages in the Empirical Programme of Relativism." *Social Studies of Science* 11 (1): 3–10. https://doi.org/10.1177/030631278101100101.

Currie, Adrian. 2018. *Rock, Bone, and Ruin: An Optimist's Guide to the Historical Sciences.* Cambridge, MA and London: The MIT Press. https://doi.org/10.7551/mitpress/11421.001.0001.

Dewey, John. 1910. *The Influence of Darwin on Philosophy, and Other Essays in Contemporary Thought.* New York: Henry Holt & Co.

Dewey, John. [1910] 1933. *How We Think: A Restatement of the Relation of Reflective Thinking to the Educative Process.* Revised edition. Lexington: D. C. Heath and Company.

Dewey, John. 1916a. *Democracy and Education: An Introduction to the Philosophy of Education.* New York: The Macmillan Company.

Dewey, John. 1916b. *Essays in Experimental Logic.* New York: Dover.

Dewey, John. 1920. *Reconstruction in Philosophy.* New York: Henry Holt and Company.

Dewey, John. 1922. *Human Nature and Conduct: An Introduction to Social Psychology.* New York: Henry Holt and Company.

Dewey, John. [1925] 1929a. *Experience and Nature.* 2nd revised edition. London: George Allen & Unwin, Ltd.

Dewey, John. 1929b. *The Quest for Certainty: A Study of the Relation of Knowledge and Action.* New York: Minton, Balch & Company.

Dewey, John. [1934] 1980. *Art as Experience.* New York: Perigee.

Dewey, John. 1938. *Logic: The Theory of Inquiry.* New York: Henry Holt and Company.

Dewey, John. 1941. "Propositions, Warranted Assertibility, and Truth." *The Journal of Philosophy* 38 (7): 169–86.

Diels, Hermann, and Walther Kranz. 1960. *Die Fragmente der Vorsokratiker; Griechisch und Deutsch. Erster Band.* Neunte Auflage. Berlin: Weidmannsche Verlagsbuchhandlung.

Diogenes Laërtios. 1905. *The Lives and Opinions of Eminent Philosophers.* English translation by C. D. Yonge. London: George Bell and Sons.

Duhem, Pierre. [1906] 1954. *The Aim and Structure of Physical Theory.* Princeton: Princeton University Press.

van Fraassen, Bas. 1980. *The Scientific Image.* Oxford: Clarendon Press. https://doi.org/10.1093/0198244274.001.0001.

van Fraassen, Bas. 1989. *Laws and Symmetry.* Oxford: Clarendon Press. https://doi.org/10.1093/0198248601.001.0001.

Fuller, Steve. 1992. "Social Epistemology and the Research Agenda of Science Studies." In *Science as Practice and Culture,* edited by Andrew Pickering, 390–428. Chicago, IL: University of Chicago Press.

Gibson, James J. 1979. *The Ecological Approach to Visual Perception.* Hillsdale, NJ: Lawrence Erlbaum.

Giere, Ronald N. 2006. *Scientific Perspectivism.* Chicago, IL and London: The University of Chicago Press. https://doi.org/10.7208/chicago/9780226292144.001.0001.

Gooding, David. 1990. *Experiment and the Making of Meaning: Human Agency in Scientific Observation and Experiment.* Boston, MA: Kluwer. https://doi.org/10.1007/978-94-009-0707-2.

Gooding, David, Trevor Pinch, and Simon Schaffer, eds. 1989. *The Uses of Experiment.* Cambridge: Cambridge University Press.

Hacking, Ian. [1983] 2010. *Representing and Intervening.* Cambridge: Cambridge University Press. https://doi.org/10.1017/CBO9780511814563.

Heidegger, Martin. [1927] 1977. *Sein und Zeit,* edited by F.-W. von Herrmann. Gesamtausgabe, Band 2. Frankfurt am Main: Vittorio Klostermann.

Hintikka, Jaakko. 1969. *Tieto on valtaa ja muita aatehistoriallisia esseitä (Knowledge Is Power, and Other Essays in the History of Ideas.)* Porvoo: WSOY.

Husserl, Edmund. [1900] 1975. *Logische Untersuchungen. Erster Band: Prolegomena zur reinen Logik. Husserliana Band XVIII,* edited by E. Holenstein. The Hague: Martinus Nijhoff.

Husserl, Edmund. [1901] 1984a. *Logische Untersuchungen. Zweiter Band: Erster Teil: Untersuchungen zur Phänomenologie und Theorie der Erkenntnis. Husserliana Band XIX/1,* edited by U. Panzer. Dordrecht: Springer.

Husserl, Edmund. [1901] 1984b. *Logische Untersuchungen. Zweiter Band: Zweiter Teil: Untersuchungen zur Phänomenologie und Theorie der Erkenntnis. Husserliana Band XIX/2,* edited by U. Panzer. Dordrecht: Springer.

Husserl, Edmund. [1936] 1976a. *Krisis der europäischen Wissenschaften und die transzendentale Phänomenologie: Eine Einleitung in die phänomenologische Philosophie, herausgegeben von Walter Biemel. Husserliana Band VI,* 2nd edition. The Hague: Martinus Nijhoff.

Husserl, Edmund. [1939] 1976b. "Die Frage nach dem Ursprung der Geometrie als intentionalhistorisches Problem." In *Die Krisis der europäischen Wissenschaften und die transzendentale Phänomenologie: Eine Einleitung in die phänomenologische Philosophie, herausgegeben von Walter Biemel. Husserliana Band VI,* 2nd edition, edited by Eugen Fink, 365–86. The Hague: Martinus Nijhoff (Originally written in 1936. First published as Fink, Eugen. 1939. *Revue Internationale de Philosophie* 1 (2): 203–25).

Husserl, Edmund. 2002. *Logische Untersuchungen. Ergänzungsband. Erster Teil. Entwürfe zur Umarbeitung der VI. Untersuchung und zur Vorrede für die Neuafulage der Logischen Untersuchungen. Husserliana Band XX/1,* edited by Ullrich Melle. Dordrecht: Springer.

Husserl, Edmund. 2005. *Logische Untersuchungen. Ergänzungsband. Zweiter Teil. Texte für die Neufassung der VI. Untersuchung: Zur Phänomenologie des Ausdrucks und der Erkenntnis. Husserliana Band XX/2,* edited by Ullrich Melle. Dordrecht: Springer.

Ihde, Don. 1979. *Technics and Praxis: A Philosophy of Technology.* Dordrecht, Boston, MA, and London: D. Reidel. https://doi.org/10.1007/978-94-009-9900-8.

Ihde, Don. 1990. *Technology and the Lifeworld: From Garden to Earth.* Bloomington and Indianapolis, IN: Indiana University Press.

Ihde, Don. 1991. *Instrumental Realism: The Interface between Philosophy of Science and Philosophy of Technology.* Bloomington and Indianapolis: Indiana University Press.

Ihde, Don. 1998. *Expanding Hermeneutics: Visualism in Science.* Evanston, IL: Northwestern University Press.

James, William. 1890. *The Principles of Psychology.* Cambridge, MA: Harvard University Press.

James, William. [1897] 1907. *The Will to Believe, and Other Essays in Popular Philosophy.* New York, London, and Bombay: Longmans Green and Co.

James, William. 1902. *The Varieties of Religious Experience.* Cambridge, MA: Harvard University Press.

James, William. [1907] 1916. *Pragmatism: A New Name for Some Old Ways of Thinking.* New York, London, Bombay, and Calcutta: Longmans, Green, and Co.

James, William. 1909. *The Meaning of Truth: A Sequel to "Pragmatism."* London, New York, Bombay, and Calcutta: Longmans, Green, and Co.

James, William. 1912. *Essays in Radical Empiricism.* New York, London, Bombay, and Calcutta: Longmans, Green, and Co.

Kant, Immanuel. [1781/87] 1956. *Kritik der reinen Vernunft.* Hamburg: Felix Meiner.

Kitcher, Philip. 1993. *The Advancement of Science: Science Without Legend, Objectivity Without Illusions.* Oxford: Oxford University Press.

Kitcher, Philip. 2001. "Real Realism: The Galilean Strategy." *The Philosophical Review* 110 (2): 151–97. https://doi.org/10.1215/00318108-110-2-151.

Knorr-Cetina, Karin. 1981. *The Manufacture of Knowledge: An Essay on the Constructivist and Contextual Nature of Science.* Oxford: Pergamon Press.

Kuhn, Thomas S. [1962] 1996. *The Structure of Scientific Revolutions.* Chicago, IL: University of Chicago Press.

Kuusinen, Otto Wille, ed. 1959. *Osnovy marksizma-leninizma. Uchebnoe posobie (The Fundamentals of Marxism-Leninism. A Textbook).* Moscow: Gosudarstvennoe izdatelstvo politicheskoy literatury.

Lakatos, Imre. 1978a. *The Methodology of Scientific Research Programmes: Philosophical Papers Volume 1.* Cambridge: Cambridge University Press.

Lakatos, Imre. 1978b. *Mathematics, Science and Epistemology: Philosophical Papers Volume 2.* Cambridge: Cambridge University Press.

Lakoff, George, and Mark Johnson. 1999. *Philosophy in the Flesh: The Embodied Mind and Its Challenge to Western Thought.* New York: Basic Books.

Latour, Bruno. 1987. *Science in Action: How to Follow Scientists and Engineers Through Society.* Cambridge, MA: Harvard University Press.

Latour, Bruno, and Steve Woolgar. [1979] 1986. *Laboratory Life: The Construction of Scientific Facts.* 2nd edition. Princeton: Princeton University Press.

Laudan, Larry. 1977. *Progress and Its Problems: Towards a Theory of Scientific Growth.* Berkeley, CA: University of California Press.

Laudan, Larry. 1981. "A Confutation of Convergent Realism." *Philosophy of Science* 48 (1): 19–49.

Laudan, Larry. 1984. *Science and Values.* Berkeley, CA: University of California Press.

Laudan, Larry. 1990. "Demystifying Underdetermination." In *Scientific Theories.* Minnesota Studies in the Philosophy of Science, vol. 14, edited by C. Wade Savage, 267–97. Minneapolis, MN: University of Minnesota Press.

Lindholm, Juho. 2021. "Science as Art: A Deweyan Perspective." In *Yearbook of the Irish Philosophical Society 2019/20. Special Issue: Science and Politics*, edited by Cara Nine. Accessed January 2, 2023. http://www.irish-philosophical-society.ie/wp-content/uploads/2021/06/Lindholm_Science-as-Art_IPS-2021-Final.pdf.

Lindholm, Juho. 2022. "The Pragmatist Basis of Material Hermeneutics." *World Academy of Science, Engineering and Technology International Journal of Humanities and Social Sciences* 16 (11). Accessed January 12, 2023. https://publications.waset.org/10012795/the-pragmatist-basis-of-material-hermeneutics.

Lindholm, Juho. 2023a. "Cybernetic Epistemology." *Acta Baltica Historiae et Philosophiae Scientiarum* 11 (1): 3–51. https://doi.org/10.11590/abhps.2023.1.01.

Lindholm, Juho. 2023b. "Scientific Practices as Social Knowledge." *International Studies in the Philosophy of Science* 35 (3–4): 223–42. https://doi.org/10.1080/02698595.2023.2196930.

Lindholm, Juho. Forthcoming. "Technics as Hermeneutics." *InfTars—Információs Társadalom* XXIII (2). Accessed July 28, 2023. https://inftars.infonia.hu/issues.

Lõhkivi, Endla, and Rein Vihalemm. 2012. "Guest Editorial: Philosophy of Science in Practice and Practical Realism." *Studia Philosophica Estonica* 5 (2): 1–6. https://doi.org/10.12697/spe.2012.5.2.01.

Määttänen, Pentti. 2009. *Toiminta ja kokemus: Pragmatistista terveen järjen filosofiaa* (Action and Experience: Pragmatist Common Sense Philosophy). Helsinki: Gaudeamus.

Määttänen, Pentti. 2015. *Mind in Action: Experience and Embodied Cognition in Pragmatism*. Cham, Heidelberg, New York, Dordrecht, and London: Springer. https://doi.org/10.1007/978-3-319-17623-9.

Marx, Karl, and Friedrich Engels (MEW). 1956–. *Karl Marx Friedrich Engels Werke*. 46 volumes. Berlin: Dietz Verlag.

Menary, Richard. 2010. "Introduction to the Special Issue on 4E Cognition." *Phenomenology and the Cognitive Sciences* 9 (4): 459–63. https://doi.org/10.1007/s11097-010-9187-6.

Merleau-Ponty, Maurice. [1942] 1967. *The Structure of Behavior (La structure du comportement)*. English translation by Alden L. Fisher. Boston, MA: Beacon Press.

Merleau-Ponty, Maurice. [1945] 2002. *Phenomenology of Perception (Phénoménologie de la perception)*. English translation by Colin Smith. London and New York: Routledge.

Newen, Albert, Leon de Bruin, and Shaun Gallagher, eds. 2018. *The Oxford Handbook of 4E Cognition*. Oxford: Oxford University Press. https://doi.org/10.1093/oxfordhb/9780198735410.001.0001.

Niiniluoto, Ilkka. 1987. *Truthlikeness*. Dordrecht: Reidel. https://doi.org/10.1007/978-94-009-3739-0.

Niiniluoto, Ilkka. 1999. *Critical Scientific Realism*. Oxford and New York: Oxford University Press.

Niiniluoto, Ilkka. 2018. *Truth-Seeking by Abduction*. Cham: Springer. https://doi.org/10.1007/978-3-319-99157-3.

Papineau, David. 2010. "Realism, Ramsey Sentences and the Pessimistic Meta-Induction." *Studies in History and Philosophy of Science* 41 (4): 375–85. https:// doi.org/10.1016/j.shpsa.2010.10.002.

Pasnau, Robert. 1997. *Theories of Cognition in the Later Middle Ages*. Cambridge: Cambridge University Press.

Peirce, Charles Sanders (CP). 1931–58. *The Collected Papers of Charles Sanders Peirce*, edited by C. Hartshorne, P. Weiss, and A. W. Burks. 8 volumes. Vols. 1–6 edited by Charles Hartshorne and Paul Weiss; Vols. 7–8 edited by Arthur W. Burks. Cambridge, MA: Harvard University Press.

Peirce, Charles Sanders (EP). 1992–98. *The Essential Peirce: Selected Philosophical Writings*, edited by N. Houser, C. J. W. Kloesel, and The Peirce Edition Project. 2 volumes. Vol. 1 edited by Nathan Houser and Christian J. W. Kloesel; Vol. 2 edited by The Peirce Edition Project. Bloomington, IN: Indiana University Press.

Peirce, Charles Sanders (MS). 1966–67. "Manuscripts." In *Charles Sanders Peirce Papers* (MS Am 1632). Cambridge, MA: Houghton Library, Harvard University. 1966: *The Charles S. Peirce Papers, Microfilm Edition, Thirty Reels with Two Supplementary Reels Later Added*. Cambridge, MA: Harvard University Library Photographic Service. Cataloged in Robin, Richard S. 1967. *Annotated Catalogue of the Papers of Charles S. Peirce*. Amherst: University of Massachusetts Press. Accessed January 3, 2023. https://hollisarchives.lib.harvard.edu/repositories/24/ resources/6437.

Pickering, Andrew. 1984. *Constructing Quarks: A Sociological History of Particle Physics*. Chicago, IL: University of Chicago Press.

Pickering, Andrew. 1995. *The Mangle of Practice: Time, Agency, and Science*. Chicago, IL: University of Chicago Press. https://doi.org/10.7208/chicago /9780226668253.001.0001.

Plato. 1952. *Meno*. In *Plato in Twelve Volumes IV*. Greek–English edition. Translated by W. R. M. Lamb, edited by T. E. Page, E. Capps, W. H. D. Rouse, L. A. Post, and E. H. Warmington, 259–371. Cambridge, MA and London: Harvard University Press and William Heinemann.

Plato. 1977. *Theaetetus*. In *Plato in Twelve Volumes VII*. Greek–English edition. Translated by H. N. Fowler, edited by G. P. Goold, 1–257. Cambridge, MA and London: Harvard University Press and William Heinemann.

Polanyi, Michael. [1958] 1962. *Personal Knowledge: Towards a Post-Critical Philosophy*. London: Routledge.

Polanyi, Michael. [1966] 1983. *The Tacit Dimension*. Gloucester, MA: Peter Smith.

Popper, Karl Raimund. [1934] 1953. *The Logic of Scientific Discovery*. New York: Harper and Row.

Psillos, Stathis. 1999. *Scientific Realism: How Science Tracks Truth*. New York and London: Routledge.

Putnam, Hilary. 1981. *Reason, Truth and History*. Cambridge: Cambridge University Press. https://doi.org/10.1017/CBO9780511625398.

Quine, Willard van Orman. 1951. "Two Dogmas of Empiricism." *The Philosophical Review* 60 (1): 20–43. https://doi.org/10.2307/2181906.

Quine, Willard van Orman. 1995. *From Stimulus to Science.* Cambridge, MA: Harvard University Press.

Radder, Hans. 1988. *The Materialization of Science.* Assen: Van Gorcum.

Rheinberger, Hans-Jörg. 1992. *Experiment, Differenz, Schrift.* Marburg: Basiliskenpresse.

Rouse, Joseph. 1987. *Knowledge and Power: Toward a Political Philosophy of Science.* Ithaca, NY and London: Cornell University Press.

Rouse, Joseph. 1996. *Engaging Science: How to Understand Its Practices Philosophically.* Ithaca, NY and London: Cornell University Press. https://doi.org/10.7591/9781501718625.

Rouse, Joseph. 2002. *How Scientific Practices Matter: Reclaiming Philosophical Naturalism.* Chicago, IL and London: The University of Chicago Press.

Russell, Bertrand. [1918] 2010. *The Philosophy of Logical Atomism.* London and New York: Routledge.

Ryle, Gilbert. [1949] 1951. *The Concept of Mind.* Watford: William Brendon and Son, Ltd., and The Mayflower Press.

Short, Thomas L. 2007. *Peirce's Theory of Signs.* Cambridge: Cambridge University Press.

Sosa, Ernest, Jaegwon Kim, Jeremy Fantl, and Matthew McGrath, eds. 2008. *Epistemology: An Anthology.* 2nd edition. Malden, Oxford and Victoria: Blackwell.

Stanford, P. Kyle. 2006. *Exceeding Our Grasp: Science, History, and the Problem of Unconceived Alternatives.* Oxford: Oxford University Press. https://doi.org/10.1093/0195174089.001.0001.

Traweek, Sharon. 1988. *Beamtimes and Lifetimes.* Cambridge, MA: Harvard University Press.

Vehkavaara, Tommi. 2007. "From the Logic of Science to the Logic of the Living. The Relevance of Charles Peirce to Biosemiotics." In *Introduction to Biosemiotics: The New Biological Synthesis*, edited by M. Barbieri, 257–98. Cham: Springer. https://doi.org/10.1007/1-4020-4814-9_11.

Vihalemm, Rein. 2001. "Chemistry as an Interesting Subject for the Philosophy of Science." In *Estonian Studies in the History and Philosophy of Science.* Boston Studies in the Philosophy of Science, vol. 219, edited by Rein Vihalemm, 185–200. Dordrecht: Kluwer. https://doi.org/10.1007/978-94-010-0672-9_14.

Vihalemm, Rein. 2011. "Towards a Practical Realist Philosophy of Science." *Baltic Journal of European Studies* 1 (1): 46–60.

Vihalemm, Rein. 2012. "Practical Realism: Against Standard Scientific Realism and Anti-Realism." *Studia Philosophica Estonica* 5 (2): 7–22. https://doi.org/10.12697/spe.2012.5.2.02.

Vihalemm, Rein. 2013. "What Is a Scientific Concept: Some Considerations Concerning Chemistry in Practical Realist Philosophy of Science." In *The Philosophy of Chemistry: Practices, Methodologies and Concepts*, edited by Jean-Pierre L\lored, 364–84. Cambridge: Cambridge Scholars Publishing.

Vihalemm, Rein. 2015. "Philosophy of Chemistry against Standard Scientific Realism and Anti-Realism." *Philosophia Scientiæ* 19 (1): 99–113. https://doi.org/10.4000/philosophiascientiae.1055.

Whitehead, Alfred North. [1929] 1978. *Process and Reality*. New York: Macmillan Publishing.

Wittgenstein, Ludwig. 1922. *Tractatus Logico–Philosophicus*. London: Routledge and Kegan Paul.

Wittgenstein, Ludwig. [1953] 2009. *Philosophical Investigations*, edited by G. E. M. Anscombe, P. M. S. Hacker, and J. Schulte. 4th revised edition. Chichester: Wiley-Blackwell.

Wittgenstein, Ludwig. 1969. *On Certainty*, edited by G. E. M. Anscombe and G. H. von Wright. New York: Harper and Row.

von Wright, Georg Henrik. 1963. *Norm and Action: A Logical Enquiry*. London: Routledge and Kegan Paul.

Chapter 2

Vihalemm's Practical Realism

A Wittgensteinian Appraisal

David Hommen

In the chorus of the increasingly numerous promoters and adherents of the so-called practical turn in philosophy of science, Rein Vihalemm's voice stands out unmistakably.[1] What distinguishes his contribution to this movement from others is its equally unique and ingenious combination of two traditions: that which emerged from the reception of Thomas S. Kuhn's (1996) *The Structure of Scientific Revolutions* on the one hand, and that of historical materialism in the wake of Karl Marx's "Theses on Feuerbach" (cf. Marx and Engels 1998, 569–71) on the other. The *uniqueness* of this confluence is perhaps explained geohistorically by the influence of both Soviet Marxist philosophy (cf. Vihalemm 2016) and the Finnish school of philosophy of science (cf. Niiniluoto 1993), to which Vihalemm has been exposed in his native Estonia and his place of work, Tartu. The *ingenuity* of Vihalemm's synthesis of Kuhn and Marx, however, lies in its claim to provide a realist philosophy of science beyond (what are considered untenable alternatives of) standard scientific realism and antirealism—a *practical* realist philosophy of science, that is.

In formulating and developing his "practical realism" (Vihalemm 2005, 180), Vihalemm is not only keenly aware of the shoulders on which he stands. He sharpens his own position in particular through comparisons and critical engagements with other practice-oriented approaches in the philosophy of science, including Sami Pihlström's (2009, 2012) pragmatic realism, Ilkka Niiniluoto's (1999) critical scientific realism, and Joseph Rouse's (1987) radical philosophical naturalism (cf. Vihalemm 2011, 2012). To these comparative discussions I would like to add another one in this chapter, focusing on a philosopher to whom Vihalemm does not explicitly refer in his writings, but whose philosophy may have had at least an indirect impact on Vihalemm's views via the work of Kuhn: Ludwig Wittgenstein. Just as

Kuhn did in philosophy of science, Wittgenstein previously initiated a practical turn in philosophy of language and cognition. And although Wittgenstein himself, of course, has no theory of science (and his position is sometimes even declared to be antiscientistic), one can nevertheless extrapolate from his linguistic-cognitive investigations a conception of science that bears a striking resemblance to Kuhn's paradigm model (cf. Kindi 1995). Indeed, Kuhn explicitly acknowledges Wittgenstein's influence on his work, drawing on such Wittgensteinian notions as family resemblance (cf. Kuhn 1996, 45) and aspect perception (cf. Kuhn 1996, 114) to explain his ideas of paradigms and paradigm shifts. In essence, then, Kuhn's *The Structure of Scientific Revolutions* can be seen as a Wittgensteinian theory of science, so that the practical turn in philosophy of science that was also taken and continued by Vihalemm may, in fact, have been originally prompted by Wittgenstein's ideas on mind and meaning (cf. Collin 2011, 26).

Furthermore, although Vihalemm approaches the problem of realism from the perspective of philosophy of science and especially philosophy of chemistry, he does not limit himself to the *specific* issues of scientific realism such as the question of the ontological status of unobservables or the semantic interpretation of theoretical terms. Rather, he is concerned with no less than our fundamental epistemic access to reality and the fundamental nature of that reality. His professed goal is to show that "[k]nowledge, the knower, and the world which is known, are all formed in and through practice," and that "science as practice is a way [i.e., *one* way—D. H.] of engaging with the world that allows the world to show how it can be identified in some of its possible 'versions'" (Vihalemm 2015, 102). Its intended breadth and depth thus bring Vihalemm's practical realism closer in spirit to Wittgenstein's practice-based philosophy than its original scientific context might at first suggest.

In view of these considerations, then, a comparison of Vihalemm's and Wittgenstein's philosophical views seems quite legitimate, both historically and systematically. The aim of the comparison sought here, however, is not merely doxographic; nor is it merely to further demarcate Vihalemm's Kuhnian-Marxist theory of science. Rather, it is to help identify and, if possible, remedy some points in which I believe his practical realism proves deficient or at least in need of supplementation. Thus, I shall argue in the following that, contrary to his own ambitions, Vihalemm falls short of resolving the dispute between (scientific) realists and antirealists; for his position effectively equals in one significant respect the view of one of the disputing parties, namely the antirealists. In contrast, it will be shown that Wittgenstein's remarks on linguistic-cognitive practices and the conditions of their possibility already supply all the resources necessary for the construction of a practical realist theory of science worthy of the name. This also gives me the opportunity to clarify the Wittgensteinian standpoint in the realism debate, which I think is all too often misunderstood. Thus, I hope that the following comparison of Vihalemm's and Wittgenstein's views will be fruitful for a better understanding of both positions, as well as for the general issue of (scientific) realism.

VIHALEMM'S PRACTICAL REALISM

Vihalemm characterizes his practical realism by the following five theses (Vihalemm 2011, 48; cf. Vihalemm 2005, 180–81; 2015, 102; Lõhkivi and Vihalemm 2012, 3):

1. Science does not represent the world "as it really is" from a God's-eye point of view.
2. The fact that the world is not accessible independently of theories—or, to be more precise, paradigms (practices)—developed by scientists does not mean that Putnam's internal realism (or social constructivism) is acceptable.
3. Science as a theoretical activity is only one aspect of it (of science) as a practical activity whose main form is scientific experimentation which in its turn takes place in the real world, being a purposeful and critically theory-guided constructive, manipulative, material interference with nature.
4. Science as practice is also a social-historical activity, which means, among other things, that scientific practice includes a normative aspect, too, and that means, in turn, that the world as it is actually accessible to science is not free from norms either.
5. Though neither naïve nor metaphysical, it is certainly realism as it claims that what is "given" in the form of scientific practice is an aspect of the real world.

With each of these theses, Vihalemm delineates his theory from other approaches in the debate on scientific realism. I would now like to discuss some of them in more detail in order to show where, from my point of view, the peculiarities, some ambiguities, and also limitations of Vihalemm's position are to be found.

With thesis (1), Vihalemm positions himself against standard scientific realism. According to him, the latter is defined by the following four propositions (Vihalemm 2012, 10):

i. There is a mind-independent world (reality) of observable and unobservable objects (the metaphysical-ontological aspect).

ii. The central notion is truth as correspondence between scientific statements (theories) and reality (the semantic aspect).
iii. It is possible to obtain knowledge about mind-independent reality (the
epistemological aspect).
iv. Truth is an essential aim of scientific inquiry (the methodological
aspect).

According to a well-known argument against standard scientific realism
(cf. Putnam 1981, 49), which Vihalemm adopts here, propositions (i), (ii),
and (iii) together imply the possibility of a so-called God's-eye point of view.
For the possibility of knowledge about a mind-independent reality implies
the possibility of recognizing the truth of (scientific) statements about this
reality in the sense of recognizing their correspondence with the latter. This,
however, presupposes the possibility of adopting a standpoint external to both
these statements and reality, from which the correspondence between the two
can be recognized. But such an external standpoint is not possible. Therefore,
statements (i) through (iii) cannot be true together, and standard scientific
realism is false.

However, as Vihalemm immediately declares in thesis (2), classical antirealist positions—such as Putnam's internal realism or social constructivism—are not viable alternatives for him either. Just why such positions are
not "acceptable" to him is not immediately clear. At times he calls them
counterintuitive and potentially self-contradictory (cf. Lõhkivi and Vihalemm
2012, 3). More specific criticisms are, first, that they (and here Vihalemm also
includes empiricist instrumentalism) "operate in the context of traditional
philosophy of science centered on language and logic, and are not founded
on actual scientific practices" (Vihalemm 2015, 101; cf. Vihalemm 2012, 11)
and, second, that they (here Vihalemm addresses Putnam's internal realism
in particular) "[belong] to the tradition of Kantianism and cannot actually
be qualified as realism at all" (Vihalemm 2011, 54; cf. Vihalemm 2011, 49;
2012, 17). These criticisms refer to theses (3) and (5) of Vihalemm's practical
realism, respectively, which in turn are important for his discussion of Niiniluoto's and Pihlström's approaches. I will therefore take up these criticisms
again in the context of the discussion of those theses.

Before that, however, the question remains whether a realist theory of
science is actually (but in a self-refuting manner) committed to the possibility of a God's-eye perspective. In fact, Vihalemm's thesis (2) can also be
understood as denying this commitment. But then it must be explained how
a realist epistemology can be had without it. Vihalemm, for his part, opts to
tackle the concept of truth. More precisely, following Rouse (1987, 147), he
wants to replace the correspondence theory of truth as posited in proposition
(ii) of standard scientific realism with a deflationary account, according to

which both the meanings of statements and the determinations of the reality that these statements are about depend on the same conditions, namely the meaningful practical interactions of cognitive-scientific agents with said reality (cf. Vihalemm 2011, 55; 2012, 19). In this view, then, the question of the correspondence or systematic connection of these statements and determinations does not arise at all. One wonders, however, whether with such a deflationary account of truth one does not also renounce the core proposition not only of standard scientific realism but of every realist epistemology, namely, the proposition (i) that there is a mind-independent reality. Vihalemm does not seem to think so. He refers to Rouse, who writes:

> This configuration of practices (including, of course, linguistic practice) allows things to show themselves as they are in a variety of respects. . . . The recognition that the possible ways a thing can be depends upon the configuration of practices within which they become manifest should therefore not reinforce the realist's fear that we are being described as "world makers." . . . The practices that constitute our "world" . . . do not determine which things exist, with what properties. (Rouse 1987, 160–61)

But what things can there be, and what properties can they have, if according to the deflationary theory of truth, all determinations of reality are conditioned by our interactions with it? The "reality-as-it-is," which is supposed to show itself in these practices, appears to be an empty concept. Vihalemm himself says:

> I take [the world itself] to be unidentified objective reality or matter, objective in the absolute sense, i.e., independent from anyone's mind or consciousness; this absolute objectivity of its existence is its only defining characteristic, it is "matter as such." It was the "thing-in-itself" for Kant; however, for practical realists or materialists it is not ungraspable, but identifiable in its concrete forms of existence through practice, being itself a concrete way of objective existence. (Vihalemm 2012, 19)

To be sure, Vihalemm may say that our "world-versions" are still versions of the world (cf. Vihalemm 2012, 17), in the sense that the world (and not we) is the *subject* of the determinations that emerge from our interaction with it. But this does not change the fact that the *source* of these determinations, according to his (and Rouse's) view, is our practices, not the world itself. In this context, Vihalemm occasionally seems to trade on an ambiguity of the term "identify," which oscillates between an epistemological sense (in which the—antecedently existing—forms of reality are merely registered in the identificatory process) and a constructivist sense (in which these forms are actively created in the very act of identification). Vihalemm undoubtedly understands the term in the latter, constructivist sense, as is clear from the cookie-cutter metaphor he employs (cf. Vihalemm 2012, 18). Thus, according to him, we identify the forms of reality in essentially the same way that we identify cookie forms in a cookie dough: by forming them ourselves (cutting them out, to continue the metaphor). But then, how can Vihalemm accommodate the idea of a mind-independent reality in his practical realism?

The main thrust of Vihalemm's argument for realism—also in the passage quoted above—seems to rest entirely on the notion of practice and, in particular, its "objective existence." So, it is time to turn to this concept. In thesis (3), Vihalemm lays emphasis on the practical nature of science as a "purposeful and critically theory-guided constructive, manipulative, material interference with nature" (Vihalemm 2011, 48); and he criticizes both standard scientific realism and (as already mentioned) its classical antirealist rivals for not giving due credit to this practical nature, focusing instead on an idea of science according to which its development "lies in constantly discovering new facts about the world and, by creating theories, connecting these facts in a logical manner, achieving a more complete and exact knowledge" (Vihalemm 2011, 47). This accusation he also levels against Niiniluoto and his critical scientific realism (cf. Niiniluoto 1999), for which he otherwise professes some sympathy (cf. Vihalemm 2011, 54; 2012, 18).

But what is the point of this turn from scientific theory to scientific practice? One might be to stress that scientific theories ultimately *are* (realized in) practices of a certain kind, for example, practices of describing, explaining, and predicting the phenomena of the empirical world. But Vihalemm obviously wants to say more. As he emphasizes, "theoretical activity is only one aspect of science"—its "main form is scientific experiment" (Vihalemm 2015, 102). Now, as Vihalemm has noted before (cf. Vihalemm 2001, 189), one characteristic of any experimental science is its constructive basis. Here he cites chemistry as a particularly salient example: this is poietic in the sense that it "is constantly enlarging the world it studies by making new stuffs" (van Brakel 1999, 134). However, according to Vihalemm, this technological or artisanal aspect is inherent to all science. Just as "[t]he aim of chemistry . . . is . . . producing substances with certain properties by means of transforming them, and discovering laws of nature about these substances," science in general "does not describe the 'given' reality 'as it is,' but does it only from the aspect of the laws of nature, constructing idealisations for this, which model the reality from the viewpoint of the technological practicability of these idealisations" (Vihalemm 2001, 190).

It is not clear how this constructivist, poietic, or technological dimension of scientific practice is supposed to contribute to a realist understanding of science. If anything, it rather seems to contradict another tenet of standard scientific realism, expressed in proposition (iv): that an essential aim of scientific inquiry is truth. But Vihalemm's reference to the experimental nature of scientific practice may have another intention. Perhaps Vihalemm wants to suggest that scientific experimenters, whatever their ultimate goals, have a particularly intimate, almost tactile contact with reality that theorists in the ivory tower of their abstract thinking lack. If this is Vihalemm's suggestion, however, it is questionable whether it is actually capable of establishing a realist philosophy of science. Skeptics might counter at this point that even (quasi)haptic experiences of reality achieved through active exploration rather than passive perception are in the end also only appearances, and possibly only inwardly generated ones at that. The attempt to prove realism with reference to the experimental nature of scientific practice seems at first no more convincing than Dr. Johnson's infamous attempt to refute George Berkeley's immaterialism by kicking a stone (cf. Boswell 1998, 333). On the other hand, if the emphasis on experimentation is merely meant to illustrate that scientists, in the course of their inquiries, enter into causal interactions with the objects of their investigation, then the distinction between theoretical and practical scientific activity is blurred. Scientific theorizing (describing, explaining, predicting) can readily be understood as a kind of mental experimentation (intellectual assimilation and accommodation) that has causal feedback loops with reality no less than its physical counterpart.

The true significance of the practical in science is, I think, stated by Vihalemm in thesis (4) of his practical realism, though even there only vaguely and potentially misleadingly: namely, that "science as practice . . . includes a normative aspect" (Vihalemm 2011, 48). That science is essentially normative means that it is essentially rule-governed and goal-directed. Scientists do not just react to the world in a causal-discriminatory way—rather, they adopt a particular *attitude* toward the world, which may be evident in the way they *treat* the world (as craftsmen, technicians, engineers, etc.), but also in the way they *describe* or *conceive* of it (as theorists). Crucial to its rule-governed (and not only causally determined) character is that the (theoretical or practical) attitude of scientists is at least partly volitional and autonomous, that is, self-legislative in the Kantian sense. As we shall see, this point is also immensely important for Wittgenstein's analysis of the concept of practice.

Vihalemm, however, in his defense of realism, seems to focus primarily on the material aspect of scientific practice. Here we arrive at the Marxist part of his theory. In his discussion of Pihlström's (2009, 2012) pragmatist philosophy of science, Vihalemm (2012, 12) considers the latter's objection that

a very basic transcendental issue concerning the practice-laden representability and experienceability of reality must be taken up from the perspective of practical realism, too: according to Vihalemm's practical realism, scientific objects can, after all, only be identified within scientific practices. Thus, it would seem—at least this rearticulation should be available to the "Kantian pragmatist"—that practices provide transcendental (contextual) conditions for the possibility of there being scientifically representable objects at all—for us. (Pihlström 2012, 88–89)

Thus, even within the framework of practical realism it looks as if "scientific objects are not 'ready-made' prior to inquiry but rather arise out of, or are constructed and/or identified in the course of, inquiry" (Pihlström 2012, 88–89). Vihalemm (2012, 13) responds to this objection with a Marxist theory of practice drawn from Marx's "Theses on Feuerbach" (cf. Marx and Engels 1998, 569–71). In the first of these theses, Marx opposes classical materialist theories, which he criticizes for having grossly neglected humans' peculiar spontaneity and autonomy. Thus, in classical (e.g., French) materialism, humans are nothing but natural objects among natural objects (Lamettrian machines), entirely passive in the sense that they are subject only to the laws of nature and moved only by the mechanical forces of physics. In contrast, classical (e.g., German) idealism emphasizes precisely the spontaneous and autonomous side of human existence. But Marx also rejects idealism because it unacceptably dematerializes and etherealizes humans and their activity. Humans are pure spirit, and their actions only abstract thinking (or willing). Marx therefore seeks to naturalize the creative and self-determined activity of humans just as he wants to naturalize humans themselves (cf. Rotenstreich 1951, 343–51). The result is a historical materialism that regards humans as "real socio-historical being[s]" and their practical activity as a "legitimate part of objective (material) reality" (Vihalemm 2015, 107). The constructivist impact of human practice on reality thus becomes "the impact of one form of objective reality upon another—the impact of reality 'in the form of activity' on reality 'in the form of an object'" (Vihalemm 2015, 107).

One may wonder how convincing the case for Vihalemm's Marxist-materialist theory of practice actually is. In particular, one may ask why, in fact, an immaterialist picture of humans and human practice should be considered unacceptable. Marx's own reasons for this may have been more ethical, not to say political-ideological, in nature (think of his eleventh Feuerbach thesis; cf. Marx and Engels 1998, 571). And classical materialism does not really answer this question either; for its systematic arguments were mainly directed against the mind-body problem evoked by Cartesian dualism, which does not arise in idealism any more than in materialism (cf. Vartanian 1953, ch. 1). It remains to be specified, then, why human practices in general, and scientific practices in particular, must necessarily be material practices. Otherwise, Vihalemm's practical realism risks begging the question.

But there is a much more serious problem. Vihalemm agrees with Pihlström that his Marxist-materialist approach to the concept of practice amounts to the same kind of "practical-realistically reinterpreted Kantianism that was also reached by the pragmatists" (Vihalemm 2012, 12). Nevertheless, he believes that his approach allows him to dispel idealist concerns. In fact, he states: "I cannot agree—it seems to me even a *contradictio in adjecto*—that this practical-realistically reinterpreted Kantian transcendentalism might be in some sense still idealistic" (Vihalemm 2012, 11–12). Alas, if a Marxist theory of practice is supposed to be Vihalemm's answer to Pihlström's Kantian-transcendental objection to practical realism, then it must be said that his answer misses the objection. Rather, it appears that Vihalemm succumbs to a confusion concerning the concepts of idealism and realism. Indeed, one must bear in mind that the term "idealism" has two quite different meanings. In one sense it denotes the thesis that everything there is can—in one way or another—be reduced to immaterial entities. Theses contrary to this position, for which I shall use Berkeley's term "immaterialism" (3D, 257) for better distinguishability, are materialism on the one hand and dualism on the other. Dualism and materialism claim, contrary to immaterialism, that there is a mind-independent reality in the sense that there is a reality that is independent of *any immaterial* mind. Materialism furthermore claims (what dualism denies) that there is no reality that is not mind-independent in this sense. In another sense, the term "idealism" denotes the thesis that all things other than us depend for their being and/or nature on our cognitive activity. Two things are to be noted about this formulation: first, its indexical recourse to *us*, and second, its neutrality with respect to the materiality or immateriality of our being and cognitive activity. The antithesis to an idealism so conceived is that which in our discussion so far has been called realism. Thus, realism claims that there is a mind-independent reality in the sense that there is a reality that is independent of *our (material or immaterial)* minds.

Now it is important to see that idealism and immaterialism are two completely independent positions. This can be well illustrated by the examples of Berkeley's so-called idealism and Marx's historical materialism. As is well known, Berkeley holds that the being of all things (other than minds) consists in their being perceived (cf. PHK, § 3). (The being of minds, in turn, consists in their being perceivers.) Nevertheless, he denies that the being of these things necessarily consists in *our* perceiving them. For, on the one hand, he is convinced that some of these things (such as trees, the furniture in my office) continue to exist even when we do not perceive them, and, on the other hand, he admits that we sometimes perceive things that do not exist at all (e.g., when we dream, hallucinate, or succumb to optical illusions). In such cases, what the (real) being of things actually consists in, according to Berkeley, is their being perceived by a mind *different* from us, namely God (cf. PHK, § 48; 3D, 212, 214–15). It thus shows that Berkeley is an "idealist" only in one sense of the term, namely in the sense that his ontology admits of nothing other than immaterial minds and their perceptions. In another sense, and indeed the sense relevant to the present discussion, however, he is a realist; for he certainly postulates the existence of things whose being and nature are independent of our cognitive activities (cf. Rosenberg 1980, 62). If we now look at Marx's historical materialism, we find that, in it, it is exactly the other way around. For Marx, all things, including minds, can be reduced to material entities and their relations to each other. At the same time, the being and nature of all material things other than us depend on our (material) cognitive activities:

[M]an is not merely a natural being: he is a *human* natural being . . ., and has to confirm and manifest himself as such both in his being and in his knowing. Therefore, *human* objects [i.e., objects of human interest and knowledge—D. H.] are not natural objects as they immediately present themselves, and neither is . . . nature objectively . . . directly given in a form adequate to the *human* being. (Marx 1988, 155–56)

[N]*ature* . . ., taken abstractly, for itself—nature fixed in isolation from man—is *nothing* for man. . . . *Nature as nature* . . . is *nothing*. (Marx 1988, 165–66)

Marx thus proves to be a "realist" only in the sense of being a materialist; in the sense of realism relevant here, however, his position must be considered antirealist, that is, idealist (cf. Kolakowski 1971; Myers 1977; Norman 1982).

For this reason, invoking a Marxist theory of practice does not help Viha-lemm to refute Pihlström's objection, which targets the idealist strain in Vihalemm's theory in the sense of its concession of a dependency of scientific objects and their identification on our scientific practices. So, when Vihalemm claims in thesis (5) that "what is 'given' in the form of scientific practice is an aspect of the real world" (Vihalemm 2011, 48), he can uphold the truth of this claim only in the materialist sense of "real," but not in the properly realist sense. And so, the impression arises that Vihalemm does not really succeed in establishing a practical *realist* theory of science, his practical "realism" thus ironically not being realism at all. This does not mean, however, that the approach of practical realism is doomed once and for all, or that the concept of practice is per se unsuitable for grounding a realist philosophy of science. However, in order to make progress with this project, one had to better look for a source of inspiration other than Marx. Thus I come to Wittgenstein.

A WITTGENSTEINIAN APPRAISAL

I have already mentioned that Wittgenstein also placed the concept of prac-tice at the center of his philosophy, and that his particular turn to practice as the mode in which all language and cognition subsist may have constituted a crucial impetus for Kuhn's highly influential philosophy of science, on which Vihalemm's practical realism also builds. However, I suspect that neither Kuhn nor the generations of "practical realists" that followed him ever fully penetrated the implications of Wittgenstein's concept of practice for the philosophy of science and even epistemology as a whole. On the contrary, the extent to which *antirealist* philosophers have repeatedly appropriated Wittgenstein's philosophy suggests that the Wittgensteinian conception of linguistic-cognitive practice and its true significance for the issue of realism are still not well understood. To this unfortunate situation, it must be admitted, have certainly contributed Wittgenstein's notoriously terse and often ambiguous mode of expression, as well as the resulting ongoing controversies among his interpreters. It is worth taking a closer look at Wittgenstein's philosophy of mind and meaning, then, not only to compare and apply it to Vihalemm's theory of science, but also to understand and appreciate it in its own right.

Wittgenstein's thesis of the primacy of practice is perhaps most readily associated with his account of the meaning of a word as its use in the language (cf. PI, § 43). However, his thesis concerns much more than just linguistic semantics. For Wittgenstein, *all* cognition, whether scientific or nonscientific, conceptual or nonconceptual, is primarily and ultimately activity (action, practice):

[W]e don't want to say that meaning is a special experience, but that it isn't anything which happens, or happens to us, but something that we do. (PG, § 107)

Thinking is an activity, like calculating. (PG, § 124)

[C]oncepts . . . correspond to a particular way of dealing with situations. (RFM, VII, § 67)

[W]hoever sees a sample like *this* will in general use it in this way. (PI, § 73)

Thus, meaning, thinking, conceiving, perceiving, and so on are all things that are *done* (and not suffered). The point here is not so much that cognition is always something corporeal. Cognitive activities may typically be carried out with words or gestures, but they can also—at least occasionally—be conducted purely "in the head" (cf. BB, 6; RPP II, § 9). What matters is that, as activities, they are always something directed toward a goal or purpose (cf. LW I, § 291; RPP II, § 9; PPF, § 69) and, accordingly, something governed by rules or techniques designed to guide them toward the fulfillment of their purpose (cf. Z, § 418; RFM, I, § 133; RPP II, § 150; PPF, § 222). That cognitive acts are rule-governed does not mean that they fall under natural laws (although they do), but that there are standards against which these acts may be pitted.

Hence, thinking, perceiving, and so on are things which can be done *correctly* or *incorrectly* (truly or falsely, adequately or inadequately) (cf. PI, §§ 54, 186).

That cognitive acts *may* be assessed according to given standards is legitimated, as Kant saw, in the fact that cognitive agents *accept* these standards, that is, (in a broad sense) *choose* to follow them. Wittgenstein recognizes, however, that in order for their cognitions to be evaluated according to any standards at all, such standards have to be sufficiently independent of the cognizers and must govern their cognitive conduct independently of the cognizers' cognitions of those standards. Otherwise, cognizers could determine not only which standards they submit to, but also when they apply them correctly and incorrectly, which would deprive the idea that the cognizers are *subject* to these standards of all meaning: "One would like to say: whatever is going to seem correct to me is correct. And that only means that here we can't talk about 'correct'" (PI, § 258).

This is *one* lesson that can be drawn from Wittgenstein's famous considerations on the nature of rule-following (cf. PI, §§ 138–242). The very idea of following a rule requires the assumption of independent criteria according to which concrete applications of the rule can be evaluated. Given that cognizing is a species of rule-following, it follows that the sheer possibility of cognition presupposes the existence of *some* conditions external to the cognitive agent that determine the correctness of his or her cognitions. This is the crucial sense in which cognitive acts or practices cannot be purely "private" (cf. PI, § 202) for Wittgenstein: it is not that they themselves must necessarily be physical—but even as possibly purely mental acts, they must be related to *something* that exists and is characterized independently of them. (For a further defense of this conclusion, see Hommen 2023a.)

Controversial within the reception of Wittgenstein's rule-following considerations, however, is what *specifically* are the external correctness conditions of cognitive acts and practices. According to one interpretive tradition (cf. Kripke 1982; Malcolm 1986; McDowell 1984), the correctness of an agent's cognitions is determined by *other* cognitive agents, or more precisely, by the *agreement* of his or her cognitions with the cognitions of the others. But this raises the question of when two cognitions agree with each other. And here the answer cannot be that their agreement consists in another *cognition* that they agree because for *that* cognition to be correct it would have to agree with further cognitions, ad infinitum. So, the correctness of an agent's cognitions can at best consist in their *objectively factual* agreement with the cognitions of other agents. And this can only consist in the objective fact that the agents apply the *same* cognitive rules to the *same* kinds of things (or in the *same* kinds of circumstances). It follows that the correctness of an agent's cognitions must be determined, at least in part, by the *objects* (or *situations*) and their *features* to which these cognitions refer (cf. Pears 1988, 368–69).

Wittgenstein himself clarifies in several passages that cognitive practices are not (or not only) based on social agreement, but (also) on "general facts of nature" (PI, § 142; cf. PI, § 492; PPF, §§ 365–66; OC, §§ 505, 617): "[I]f things were quite different from what they actually are . . . our normal language-games would thereby lose their point.—The procedure of putting a lump of cheese on a balance and fixing the price by the turn of the scale would lose its point if it frequently happened that such lumps suddenly grew or shrank with no obvious cause" (PI, § 142). Cognitive practices depend on such natural constants (that objects generally keep their weight, for example) because only these can provide stable criteria for the correctness of individual cognitive acts (such as weighing). Thus: "The agreement, the harmony, between thought and reality consists in this: that if I say falsely that something is *red*, then all the same, it isn't *red*. And in this: that if I want to explain the word 'red' to someone, in the sentence 'That is not red,' I do so by pointing to something that *is* red" (PI, § 429; translation modified). In Wittgenstein's view, then, neither a solipsistic mind nor a community of minds can completely autonomously determine the correctness of their cognitions (and thus fix their contents); this requires the existence of a reality that exists and is characterized independently of *any* cognitive agent.

Up to this point I have summarized Wittgenstein's conception of cognition as far as it considers the phenomenon of cognitive agency "from the outside," determining what it is for there to be agents who act cognitively (the objective conditions of cognitive agency). However, in his investigations, Wittgenstein considers cognitive agency also "from the inside," asking what it is *for agents* to act cognitively (the subjective conditions of cognitive agency): "[H]ow can a rule teach me what I have to do at *this* point?" (PI, § 198). On the objective side, as we have seen, the possibility of cognitive agency presupposes mind-independently existing, intrinsically characterized objects or situations. On the subjective side, Wittgenstein says, it presupposes the recognizability of these objects (situations) and their features. For agents to be able to cognize certain things (believe, conceive, perceive them) is for them to be capable of applying rules to those things. To be capable of doing so, they must know, at least roughly, the external conditions for the application of these rules, and be able to recognize, at least to some extent, when those conditions are present (cf. PI, §§ 72, 208).

However, a problem arises when one conceives of this knowing and recognizing again as cognitions on the part of the agents. For, as we have seen, the conditions for the application of these rules consist, in part, in the very things the agents are supposed to become capable of cognizing. If their being known and recognized were again to hark back to cognitions on the part of the agents, this would mean that the agents' ability to cognize these things already presupposes their cognizing them, which of course cannot be. As

Wittgenstein puts it, "every interpretation hangs in the air together with what it interprets, and cannot give it any support" (PI, § 198). Wittgenstein solves this problem by conceiving of the knowledge and recognition of the application conditions of cognitive rules *not* as cognition proper but as what he calls "intransitive understanding" or "familiarity" (PG, § 37):

> "Something is familiar if I know what it is." . . . What kind of thing is "familiar-ity"? . . . I would like to say: "I see what I see." . . . I say that because I don't want to give a name to what I see . . .; . . . the criterion for recognition isn't that I name the object correctly, but that when I look at it I utter a series of sounds and have a certain experience. . . . So the multiplicity of familiarity, as I understand it, is that of feeling at home in what I see. (PG, § 115–16)

We may think of familiarity as a kind of *pre-cognition*. (I write "pre-cog-nition" to distinguish this from precognition in the sense of prescience and foreknowledge.) It is *not* activity (although it is not a state either), but pure receptivity (experience, reaction). It can (like cognition proper) have both inner and outer forms of expression or manifestation. It is not rule-governed (although it is not erratic either) and is therefore not subject to any standards by which it may be evaluated. Thus, when cognitive agents are familiar with the application conditions of their cognitive rules, they know and recognize these conditions not as something they are according to *other* criteria, but simply as what they *are* (cf. PG, § 118). They do not acquire a mediated, potentially deceptive understanding of those conditions, but become directly and unambiguously acquainted with them: "If I'm told: 'look for a red flower in this meadow and bring it to me' and then I find one—do I compare it with my memory picture of the colour red?—And must I consult yet another pic-ture to see whether the first is still correct?—In that case why should I need the first one?—I see the colour of the flower and recognize *it*" (PG, § 54).

This, then, is another lesson to be drawn from Wittgenstein's consider-ations on rule-following. Insofar as cognizing is a species of rule-following, all cognition is based on an ability to pre-cognize the external conditions for the application of the relevant rules in the sense that the cognitive agent grasps these conditions in an immediate, nondiscursive, nonreflective way. Thus, just as mind-independently existing, intrinsically characterized objects or situations are an objective transcendental condition for the possibility of cognitive agency, their pre-cognizability is a subjective transcendental condi-tion for the possibility of such agency, so that the actual existence of the latter proves the reality of both.

From Wittgenstein's rule-following considerations, one can derive a conception of scientific knowledge which agrees in many aspects with Viha-lemm's practical realism. First of all, there is the vehement insistence on the importance of practice for our epistemic faculties, which both approaches share. Wittgenstein, though not unquestionably asserting this for cognitive practices in general (cf. Baker and Hacker 1984), would certainly also agree with Vihalemm on the point that scientific practices are inherently social and, to that extent, informed by historico-cultural conventions. However, for Wittgenstein, as we have seen, the normativity of scientific practices begins at a much deeper level, namely at the level of activity—even of the individual act—itself.

There is also an interesting similarity between Wittgenstein's distinction between cognition and pre-cognition, on the one hand, and Vihalemm's (1999, 2001) distinction between "φ-science" and "non-φ-science," which is quite relevant to his practical conception of science, on the other. Vihalemm characterizes φ-science by a "constructive-hypothetico-deductive" method; its paradigm is physics. It is constructive insofar as it projects idealized models and mathematically formulated laws of nature onto its domain of investigation. Hence, "[i]t is φ-science, where we can suppose that objects must conform to our knowledge, not our knowledge to objects" (Vihalemm 2001, 188). Non-φ-science, by contrast, is characterized by a "classifying-descriptive-historical" method; its paradigm is (classical) biology. It is descriptive insofar as it endeavors to grasp and categorize its domain of investigation without prior theoretical assumptions. Hence, "[its objects are] 'given' in some way or other to the researcher by some kind of pre- or non-scientific practices" (Vihalemm 2001, 188). Now, just as Wittgenstein believes that all cognition involves pre-cognition, Vihalemm believes that all science contains both a φ-scientific and a non-φ-scientific component. The paradigm in which he believes this is most evident is chemistry (cf. Vihalemm 2001, 189). And just as Wittgenstein argues that pre-cognition forms the basis of all cognition, Vihalemm argues that all φ-science is based on non-φ-science (cf. Vihalemm 2001, 196). However, for Vihalemm, non-φ-science is still situated at a conceptual, and thus partly constructive, level. Thus, even in non-φ-science, "there are no objects and subjects of cognition 'ready-made' or 'given' by nature itself, since they both have a historico-cultural character as well" (Vihalemm 2001, 188). Wittgenstein's pre-cognition, on the other hand, is located way below the conceptual (let alone historico-cultural) level, and thus forms the basis not only of all φ-science, but also of all non-φ-science.

In addition to the commonalities just mentioned, we can now also see how a Wittgensteinian conception of cognitive practice can remedy the shortcomings of Vihalemm's practical realism identified earlier:

First, Wittgenstein's conception can show that there must be a reality with qualities independent of *us* and *our* cognitive activities. To this extent, this conception stands in clear opposition to idealist theories such as Kantian

idealism, pragmatism, Putnam's internal realism, but also Marxist materialism and social constructivism.

Second, Wittgenstein's conception shows that there must be entities with
traits independent of *all* cognitive activity. Thus, his conception is also
opposed to "idealist" theories such as Berkeley's immaterialism.

Third, Wittgenstein is able to elucidate the significance of the concept of
practice and, especially, the relevance of one component of this concept,
namely the concept of practical normativity, for the issue of realism.

Fourth, he can give clear sense to the idea that through our cognitive practices we come into direct epistemic contact with reality. Our pre-cognitive
acquaintance with the enabling conditions of our cognitive practices, precisely because it is pre-cognitive, does not involve any representations (propositional or eidetic) that might arouse skeptical doubts. It is not an arbitrary,
possibly fictitious interpretation of reality, but its authentic intuition.

With this, finally, Wittgenstein can also offer a realist alternative to the
classical correspondence theory of truth that has caused so much trouble for
standard scientific realism. According to him, our recognition of the correctness
of our cognitions (the truth of our thoughts and the veridicality of our perceptions) does not emerge from a *comparison* of our cognitions with reality that
we *make*, but rather from a *coincidence* (cf. PG, § 119) of the two that we *feel*:

> If we represent familiarity as an object's fitting into a sheath, that's not quite
> the same as our *comparing* what is seen with a copy. What we really have in
> mind is the feeling when the object slips smoothly into the contour of the sheath.
>
> [W]hat I would like to say is that the sheath in my mind is, as it were, the
> "form of imagining", so . . . that it is no longer presented to [the mind] as an
> object. But that only means: it didn't make sense to talk of a [sheath] at all.
> (PG, § 130)
>
> So I see only one thing [i.e., the real object—D. H.], not two. (PG, § 119)

The flaw in the classical correspondence theory is the assumption that recognizing the correctness of our cognitions is itself a cognitive act—which
we would then (impossibly) have to perform from a God's-eye perspective.
Wittgenstein's pivotal insight is that the recognition of the correctness of our
cognitions is first and foremost a pre-cognitive experience that we have *in the
course of* our cognitive acts. The correctness of our cognitions—if they *are*
correct—is already *given* to us in these cognitions. Their correctness is given
to us in these cognitions because *reality itself* is given to us in them: "I see
the colour of the flower and recognize *it*" (PG, § 54). It is in this sense, then,
that reality coincides with our cognitions.

To be sure, since it is pre-cognitive, our recognition of the correctness
of our cognitions is at first inarticulate. This need not mean that we cannot

attain a more reflective knowledge about the world. Such knowledge (which we strive for in science and metaphysics) would be accomplished, however, not by taking an external standpoint on our cognitions and reality, but by making explicit what is initially only implicit in our cognitions, namely the real conditions that make those cognitions possible (cf. Mounce 2005, 106).

EPILOGUE

As I hope to have shown, a Wittgensteinian conception of scientific practice goes some way toward realizing Vihalemm's ambitious vision of practical realism. Of course, follow-up questions arise that are worth pursuing on another occasion. One ties in directly with the issue of higher-level cognition just mentioned. With respect to this, it remains to be explained how—in general, but especially in science and metaphysics—the implicit foundations of our world knowledge can be made explicit at all. For given that these foundations are—as Wittgenstein's analysis suggests—*essentially* implicit, this effort seems tantamount to trying to say the unsayable: to explicate with concepts that which cannot be explicated with concepts. And would not such a verbalization just distort the peculiar mode of our tacit knowing?

Wittgenstein's key to resolving this paradox is *metaphorical redescription.* In metaphorical redescription, one transfers a term with all or at least some of its semantic associations from its original to a novel domain of application, thereby exhibiting certain analogies between both, and thus becoming aware of previously unnoticed features ("aspects"; cf. PPF, § 113) of the target domain. Basically, something like this happens every time we apply a familiar concept to a novel case (think of the concept "game"; cf. PI, §§ 66–69). Whether such a reapplication is then classified as a metaphorical redescription or rather as an extension of the original literal meaning of the concept in question (whether, for example, language is *really* a game or only in a figurative sense; cf. PI, § 7) is, for Wittgenstein, more of a pragmatic question (cf. BB, 139–40). What is crucial in either case is that the reapplication of the term remains substantially guided and constrained by its original semantic associations, with the effect that its metaphorical (or extended literal) meaning—that is, the analogies and features signified by it—cannot be explained and known independently of recourse to its (original) literal meaning (cf. PPF, §§ 274–78). In this sense, the metaphorical (and, in general, the) meaning of a concept is always to some extent *ineffable*. It shows itself in the (re)application of a term without being able to be said itself. Now, Wittgenstein develops the principle of metaphorical redescription into a whole philosophical method (which he calls "surveyable representation"; cf. PI, § 122), in which innovative metaphors ("intermediate links"; cf. PI, § 122) are employed to display

connections between phenomena that otherwise seem unconnected (and confusing), thus creating a poetic understanding of these phenomena which no
prosaic description of them could convey (cf. PI, § 122; BB, 28). In this way,
Wittgenstein's method undertakes the dialectical attempt to use the concepts
of our language to point beyond the limits of language, and to express the
inexpressible foundation of our cognitive enterprise.

As a matter of fact, science often proceeds according to a very similar
method. For one can also regard scientific theories and especially models as
(more or less sophisticated) metaphors, which explain parts and processes
of reality that are not yet understood by describing them in analogy to more
familiar ones (cf. Black 1962; Hesse 1966). Such a view not only dissolves
the irritating puzzle of the sometimes blatant idealizations made in scientific
models (which, in this view, are not to be understood literally but metaphorically, and thus may be literally false but still metaphorically true). It also
resolves the long-running dispute between scientific realists on the one hand,
who interpret the theoretical terms of scientific theories and models as des-
ignators of unobservable entities, and descriptivist empiricists on the other,
who conceive of such terms as merely elliptical descriptions of empirical
regularities. For as soon as one understands scientific theories and models as
metaphorical descriptions of reality in Wittgenstein's sense, the question of
what these theories and models *really* denote no longer arises. After all, their
descriptions are to be understood such that they express figuratively what
cannot be stated literally.

As I said, all this would have to be elaborated in more detail. (For the first
efforts in this direction, cf. Hommen 2022 and 2023b.) Meanwhile, I think
that a Wittgensteinian conception of scientific models as metaphors, with
its corollary for the interpretation of theoretical concepts, resonates quite
well with Vihalemm's view, which declares the very distinction between the
empirical world of observables and the transcendent world of unobservables
to be an artifact of the Cartesian-Humean-Kantian tradition in epistemology
(cf. Vihalemm 2011, 58). What matters in both Wittgenstein's theory of
cognition and Vihalemm's practical realism is not so much the observability
of the world as its "interactability." Whether we describe this interactable
world in terms of unobservable entities or in terms of empirical regularities
is irrelevant for the epistemological status of our theories and models. As
Vihalemm explains, "there is one [i.e., *only* one—D. H.] real world which
is complex, inexhaustible, and can manifest itself in practice in a potentially
infinite number of ways, i.e. in principle there can be an infinite number of
real 'world-versions'" (Vihalemm 2011, 58). What is decisive, as Wittgenstein would only add, is which of our numerous possible descriptions of the
world provides the best poetic expression—the "mot juste," so to speak (PPF,

§ 226)—for its essence: an essence with which we are tacitly but unfailingly acquainted in and through our cognitive and scientific practices.

NOTE

1. This research was funded by the Deutsche Forschungsgemeinschaft (DFG, German Research Foundation), no. 452319975. I would like to thank the two anonymous reviewers for their valuable comments and suggestions.

REFERENCES

Baker, Gordon P., and P. M. S. Hacker. 1984. *Scepticism, Rules and Language.* Worchester: Billing and Sons, Ltd.

Berkeley, George. 1949a. "Of the Principles of Human Knowledge: Part 1." In *The Works of George Berkeley, Bishop of Cloyne. Volume 2*, edited by Arthur A. Luce and Thomas E. Jessop, 41–113. London: Thomas Nelson and Sons.

Berkeley, George. 1949b. "Three Dialogues Between Hylas and Philonous." In *The Works of George Berkeley, Bishop of Cloyne. Volume 2*, edited by Arthur A. Luce and Thomas E. Jessop, 163–263. London: Thomas Nelson and Sons.

Black, Max. 1962. *Models and Metaphors: Studies in Language and Philosophy.* Ithaca: Cornell University Press. https://doi.org/10.7591/9781501741326.

Boswell, James. 1998. *Life of Johnson*, edited by Robert W. Chapman. Oxford: Oxford University Press.

Collin, Finn. 2011. *Science Studies as Naturalized Philosophy*. Dordrecht: Springer. https://doi.org/10.1007/978-90-481-9741-5.

Hesse, Mary B. 1966. *Models and Analogies in Science*. Notre Dame: University of Notre Dame Press.

Hommen, David. 2022. "Wittgenstein, Ordinary Language, and Poeticity." *Kriterion* 35 (4): 313–34. https://doi.org/10.1515/krt-2021-0036.

Hommen, David. 2023a. "Seizing the World: From Concepts to Reality." *Metaphysica* 24 (2): 421–44. https://doi.org/10.1515/mp-2022-0031.

Hommen, David. 2023b. "Semantic Understanding, Aspect Perception, and Metaphysics." In *100 Tractatus: Contributions of the Austrian Ludwig Wittgenstein Society, Vol. XXIX*, edited by Esther Heinrich-Ramharter, Alois Pichler, and Friedrich Stadler, 279–87. Kirchberg a.W.: ALWS.

Kindi, Vasso P. 1995. "Kuhn's 'The Structure of Scientific Revolutions' Revisited." *Journal for General Philosophy of Science* 26 (1): 75–92. https://doi.org/10.1007/BF01130927.

Kolakowski, Leszek. 1971. "Karl Marx and the Classical Definition of Truth." In *Marxism and Beyond: On Historical Understanding and Individual Responsibility*, translated by Jane Z. Peel, 59–87. London: Paladin.

Kripke, Saul A. 1982. *Wittgenstein on Rules and Private Language*. Cambridge, MA: Harvard University Press.

Kuhn, Thomas S. 1996. *The Structure of Scientific Revolutions*. 3rd edition. Chicago, IL and London: The University of Chicago Press.

Lõhkivi, Endla, and Rein Vihalemm. 2012. "Philosophy of Science in Practice and Practical Realism." *Studia Philosophica Estonica* 5 (2): 1–6. https://doi.org/10.12697/spe.2012.5.2.01.

Malcolm, Norman. 1986. *Nothing Is Hidden: Wittgenstein's Criticism of His Early Thought*. Oxford: Blackwell.

Marx, Karl. 1988. *Economic and Philosophic Manuscripts of 1844 and the Communist Manifesto*, translated by Martin Milligan. Amherst: Prometheus Books.

Marx, Karl, and Friedrich Engels. 1998. *The German Ideology: Including Theses on Feuerbach and Introduction to the Critique of Political Economy*. Amherst: Prometheus Books.

McDowell, John. 1984. "Wittgenstein on Following a Rule." *Synthese* 58 (3): 325–63. https://doi.org/10.1007/BF00485246.

Mounce, Howard O. 2005. "Wittgenstein and Classical Realism." In *Readings of Wittgenstein's On Certainty*, edited by Danièle Moyal-Sharrock and William H. Brenner, 103–21. Basingstoke and New York: Palgrave Macmillan. https://doi.org/10.1057/9780230505346_7.

Myers, David B. 1977. "Marx's Concept of Truth: A Kantian Interpretation." *Canadian Journal of Philosophy* 7 (2): 315–26. https://doi.org/10.1080/00455091.1977.10717021.

Niiniluoto, Ilkka. 1993. "Philosophy of Science in Finland: 1970–1990." *Journal for General Philosophy of Science* 24 (1): 147–67. https://doi.org/10.1007/BF00769519.

Niiniluoto, Ilkka. 1999. *Critical Scientific Realism*. New York: Oxford University Press.

Norman, Richard. 1982. "The Primacy of Practice: 'Intelligent Idealism' in Marxist Thought." *Royal Institute of Philosophy Supplements* 13: 155–79. https://doi.org/10.1017/S0957042X00001590.

Pears, David. 1988. *The False Prison: A Study of the Development of Wittgenstein's Philosophy. Volume 2*. Oxford: Oxford University Press. https://doi.org/10.1093/019824486X.001.0001.

Pihlström, Sami. 2009. *Pragmatist Metaphysics: An Essay on the Ethical Grounds of Ontology*. London and New York: Continuum International Publishing Group.

Pihlström, Sami. 2012. "Toward Pragmatically Naturalized Transcendental Philosophy of Scientific Inquiry and Pragmatic Scientific Realism." *Studia Philosophica Estonica* 5 (2): 79–94. https://doi.org/10.12697/spe.2012.5.2.06.

Putnam, Hilary. 1981. *Reason, Truth and History*. Cambridge: Cambridge University Press. https://doi.org/10.1017/CBO9780511625398.

Rosenberg, Jay F. 1980. *One World and Our Knowledge of It: The Problematic of Realism in Post-Kantian Perspective*. Dordrecht: D. Reidel Publishing Company.

Rotenstreich, Nathan. 1951. "Marx' Thesen über Feuerbach." *Archiv für Rechts- und Sozialphilosophie* 39 (3): 338–60.

Rouse, Joseph. 1987. *Knowledge and Power: Toward a Political Philosophy of Science*. Ithaca, NY: Cornell University Press.

Van Brakel, Jaap. 1999. "On the Neglect of the Philosophy of Chemistry." *Foundations of Chemistry* 1: 111–74. https://doi.org/10.1023/A:1009936404830.

Vartanian, Aram. 1953. *Diderot and Descartes: A Study of Scientific Naturalism in the Enlightenment*. Princeton, NJ: Princeton University Press. https://doi.org/10.1515/9781400877188.

Vihalemm, Rein. 1999. "Can Chemistry be Handled as its Own Type of Science?" In *Ars mutandi—Issues in Philosophy and History of Chemistry*, edited by Nikos Psarros and Kostas Gavroglu, 83–88. Leipzig: Leipziger Universitätsverlag.

Vihalemm, Rein. 2001. "Chemistry as an Interesting Subject for the Philosophy of Science." In *Estonian Studies in the History and Philosophy of Science*, Boston Studies in the Philosophy of Science, vol. 219, edited by Rein Vihalemm, 185–200. Dordrecht: Springer. https://doi.org/10.1007/978-94-010-0672-9_14.

Vihalemm, Rein. 2011. "Towards a Practical Realist Philosophy of Science." *Baltic Journal of European Studies* 1: 46–60.

Vihalemm, Rein. 2012. "Practical Realism: Against Standard Scientific Realism and Anti-Realism." *Studia Philosophica Estonica* 5 (2): 7–22. https://doi.org/10.12697/spe.2012.5.2.02.

Vihalemm, Rein. 2015. "Philosophy of Chemistry Against Standard Scientific Realism and Anti-Realism." *Philosophia Scientae* 19 (1): 99–113. https://doi.org/10.4000/philosophiascientiae.1055.

Vihalemm, Rein. 2016. "Theoretical Philosophy and Philosophy of Science in the Soviet Times: Some Remarks on the Example of Estonia, 1960–1990." *Studia Philosophica Estonica* 8 (2): 195–227. https://doi.org/10.12697/spe.2015.8.2.10.

Wittgenstein, Ludwig. 1969. *Preliminary Studies for the "Philosophical Investigations". Generally Known as The Blue and Brown Books*. 2nd edition. Oxford: Basil Blackwell.

Wittgenstein, Ludwig. 1974a. *On Certainty*, edited by G. E. M. Anscombe and Georg H. von Wright, translated by Denis Paul and G. E. M. Anscombe. Reprint with corrections and indices. Oxford: Blackwell Publishers, Ltd.

Wittgenstein, Ludwig. 1974b. *Philosophical Grammar*, edited by Rush Rhees, translated by Anthony Kenny. Oxford: Basil Blackwell.

Wittgenstein, Ludwig. 1978. *Remarks on the Foundations of Mathematics*. 3rd edition, edited by Georg H. von Wright, Rush Rhees, and G. E. M. Anscombe, translated by G. E. M. Anscombe. Oxford: Basil Blackwell.

Wittgenstein, Ludwig. 1980. *Remarks on the Philosophy of Psychology. Volume 2*, edited by Georg H. von Wright and Heikki Nyman, translated by C. Grant Luckhardt and M. A. E. Aue. Oxford: Basil Blackwell.

Wittgenstein, Ludwig. 1981. *Zettel*. 2nd edition, edited by G. E. M. Anscombe and Georg H. von Wright, translated by G. E. M. Anscombe. Oxford: Basil Blackwell.

Wittgenstein, Ludwig. 1982. *Last Writings on the Philosophy of Psychology. Volume 1*, edited by Georg H. von Wright and Heikki Nyman, translated by C. Grant Luckhardt and M. A. E. Aue. Oxford: Basil Blackwell.

Wittgenstein, Ludwig. 2009a. *Philosophical Investigations*. Revised 4th edition, edited by P. M. S. Hacker and Joachim Schulte, translated by G. E. M. Anscombe, P. M. S. Hacker, and Joachim Schulte. Chichester: Wiley-Blackwell.
Wittgenstein, Ludwig. 2009b. "Philosophy of Psychology—A Fragment." In *Philosophical Investigations*, edited by P. M. S. Hacker and Joachim Schulte, 182–243. Chichester: Wiley-Blackwell.

APPENDIX

Notes on references and abbreviations

For Berkeley's works, the following abbreviations are used:

PHK   "Of the Principles of Human Knowledge: Part 1"
3D   "Three Dialogues between Hylas and Philonous"

For Wittgenstein's works, the following abbreviations are used:

BB   *The Blue and Brown Books*
LW I   *Last Writings on the Philosophy of Psychology, Vol. 1*
OC   *On Certainty*
PG   *Philosophical Grammar*
PI   *Philosophical Investigations*
PPF   "Philosophy of Psychology—A Fragment"
RFM   *Remarks on the Foundations of Mathematics*
RPP II   *Remarks on the Philosophy of Psychology, Vol. 2*
Z   *Zettel*

All numbers are page numbers unless otherwise indicated.

Chapter 3

Rein Vihalemm's Practical Realism and the Pragmatics of Actual Scientific Inquiry

Why Scientific Realism Can Only Be Practical

Kenneth R. Westphal

This chapter on Rein Vihalemm's practical realism originates in a fortunate coincidence. During a fellowship Rein spent in the Helsinki Collegium for Advanced Studies, Sami Pihlström kindly invited me to spend a quarter there as a visiting scholar (October 2011–January 2012). I remain very grateful to Sami's and the collegium's generous support, and for introducing me to Rein. We quickly discovered we shared very substantial interests and hold strongly convergent views regarding scientific forms of real knowledge and their proper philosophical accounting. Rein kindly acknowledged some of these points in various papers.[1] Here I explicate and defend more of what we shared then and thereafter, to clarify, corroborate, and undergird Rein's practical realism.

WHAT IS PRACTICAL ABOUT "PRACTICAL REALISM"?

Rein contrasts his "practical realism" to a host of more familiar views and approaches, noting both convergences and divergences. His comparisons are informative, but philosophers may wonder, what philosophical merits might *practical* realism have? Why compare these broad issues and approaches, and introduce an unfamiliar alternative, or designation?

One clue to answering these questions is James Griffin's observation regarding moral philosophy, which holds generally:

One might succeed in making every argument that one actually deployed water-tight. But one does not usually go seriously wrong in philosophy over details of one's argument. One goes seriously wrong in the biggest things, in the things one does not even think of, in one's whole orientation. At the very best, one's orientation will allow one a glimpse of an important truth or two, but it will also certainly be responsible for one's overlooking a dozen others. In philosophy generally, . . . we are at present, and always shall be, groping in the dark simply to get a sense of some of the large contours of our subject. One's only reasonable hope is that, by groping, one will find something, and that others will take a look. (Griffin 1996, 2)

In regard to the history and philosophy of science (HPS), I examine below some indicative, chronic errors, both in philosophical details and in philosophical orientations within recent analytical philosophy of science, which indicate why practical realism provides an important reorientation for HPS.

Vihalemm derives his sense both of practice and realism from early Marx, who stressed that we human beings live within the natural, social, and historical world, grounded in nature and in our human nature, though our various individual, social, and historical productions are only possible and successful through how we, one natural (and social and historical) species, individually and collectively figure out how to solve various problems confronting us by making things, tools, languages, procedures, including the entirety of our economies, societies, and productive forms of disciplined inquiry (cf. Vihalemm 2011, 51; 2012, 206). Stated so directly, this may appear obvious and undeniable. However, philosophers have a long history and continuing habit of neglecting, evading, or denying these basic features of our humanity and our human accomplishments. Such habits of thought, acquired and perpetuated through much typical philosophical training, often generate philosophical perplexities, including persisting perplexities about human knowledge and its various objects, including within HPS. Educated in a very different (Soviet) context, Vihalemm reflected philosophically upon the very practical bases of human cognition from at least the mid-1960s (cf. Vihalemm 1966).

Specific examples of relevant perplexities are examined below; note first that central stress upon our human nature within our natural habitat, as also within our social and historical habitat(s), was central to the classical pragmatism of Peirce, Dewey, and Mead, as also to later pragmatic realists, such as C. I. Lewis (1929),[2] Ralph Sleeper, Frederick Will, and Sami Pihlström, with his magnificent first book.[3] However, the term "pragmatism" was hijacked early on; Peirce found it necessary to coin an ugly designation for his views, "pragmaticism," to dissuade further terminological meddling—though important realist aspects appeared early in James's thought and were central to his later writings (Hare and Chakrabarti 1980). To many contemporary philosophical eyes, the designation "pragmatic realism" appears oxymoronic, while "practical realism" appears puzzling, pointless, or vacuous. "Pragmatic realism" is not oxymoronic; neither is it pointless or vacuous. Why these designations are found puzzling merits critical scrutiny. Understandably, Vihalemm is as interested as Peirce in by-passing terminological squabbling about the use and abuse of the term "pragmatism"—as also the terms "realism" and "practice." However, much more is at stake than terminological proliferation. The kinds of broader, self-critical philosophical reflections to which Griffin directs our attention are central to Vihalemm's practical realism; to his use of his idealized model of quantified exact science, φ-science; and to his probing various ways in which scholars of natural sciences may (mis)understand their key questions about natural science and the point of answering those questions; these are especially evident in Vihalemm (2007).[4]

PRACTICAL REALISM AND "NATURALISM"

Central to practical realism is the robust naturalism highlighted in the second and eighth of Marx's (1845) "Theses on Feuerbach" (Vihalemm 2011, 51; 2012, 206), according to which all theoretical issues, including issues about human knowledge, must be addressed within the framework of human practices within our natural, worldly context, through which one aspect of nature—human agents—interacts with another aspect—nature as our ineluctable context of all human activity, including thinking, acting, investigating, discovering, assessing, and revising our understanding of effective forms of inquiry and of credible results of human inquiries. Inter alia, this invokes significant forms of what is now called "externalism," granting priority to what our world is, what our capacities and activities are, what features of the world we actually engage with, and how well our inquiries characterize these features. Each and all of these have priority over whatever we may think, say, or believe about each of these facets of, inter alia, our disciplined inquiries (Vihalemm 2011, 50; 2012, 206).

Why and how can such a broad, moderate externalism be informative? How can or does our natural environment and our natural endowment(s) enable or afford us sapience about any natural particulars or aspects of nature? That is a huge question; simply understanding this question and the inquiries required to answer it cogently are short-circuited by a prevalent philosophical dichotomy between "reason" and "cause," as in Davidson's view that any belief can only be justified by other beliefs, whereas beliefs are caused (not justified) by perception.[5] Regarding his clinical subject, "Schneider," who suffered profound forms of motor-visual aphasia due to a shrapnel wound to his occipital region, Merleau-Ponty observed:

His visual deficiencies are extensive, but . . . it would be absurd to explain all others through them as if through their cause, and it is no less absurd to think that the shrapnel collided with symbolic consciousness. (Merleau-Ponty 2005, 158; 2012, 127)

Originally, Davidson merely claimed that beliefs are caused by perception; later he clarified his view somewhat, stating:

To perceive that it is snowing is, under appropriate circumstances, to be caused (in the right way) by one's senses to believe that it is snowing by the actually falling snow. Sensations no doubt play their role, but that role is not that of providing evidence for the belief. (Davidson 2001, xvi, cf. 164, 174, 193–204)

For (at least) three reasons, this claim is merely promissory: a definite account of such perceptual belief requires specifying "appropriate" circumstances, what counts as "the right way" to be so caused to believe, and above all: it requires addressing one of Dretske's (1981, 1–82) key findings, that in principle causal relations neither suffice nor are necessary for semantic content (intension, meaning, classification), which requires distinct and much more stringently specified information relations, and that such intension is (centrally) constitutive of beliefs, thoughts, sentences, or assertions. Burge (1999, 249n16) expressly referred Davidson to Dretske, including *Knowledge and the Flow of Information*, though without mentioning the key finding noted here. However, I have found no published indication that Davidson considered Dretske's work. Indeed, Dretske's information-theoretic epistemology is widely assimilated to generic causal-reliability epistemology—a grotesque error, however convenient it may be to presume a simple dichotomy between "reason" and "cause," and perhaps dispense with "reasons" altogether in pursuit of a radical causal naturalism (such as Quine's). A pervasive problem with contemporary versions of causal naturalism is that they strongly tend to presuppose what counts as "causal," rather than examining the wide and important varieties of causation identified in relevant sciences. Such research is required, inter alia, to identify actual information channels involved in human sensation, perception, and cognition. Dretske wrote that book to foster multidisciplinary cooperation and inquiry, but his radical reconstruction of Shannon and Weaver's account of information quantity and its sufficiently reliable transmission into a specifically *semantic* theory of information content, though not unprecedented, was widely misunderstood by both philosophers and cognitive scientists.

That epistemology must be pursued by multidisciplinary and international cooperations was well understood at the turn of the twentieth century, when the leading journal *Mind* was founded (1876) with an overtly multidisciplinary title: *Mind: A Quarterly Review of Psychology and Philosophy*, which took "psychology" very broadly and featured even more extensively multidisciplinary contributions—for example, in ethology, neuroanatomy, comparative anatomy—by leading international scholars (cf. Westphal 2020). Such multidisciplinarity and cosmopolitan orientation soon faded as "analytical" philosophy came to dominate Anglophone philosophy. The extent to which contemporary analytic approaches to epistemology and to philosophy of mind avoid actual sciences and their findings insulates such philosophy from assessment by other disciplinarians, though at the expense of philosophical relevance to other disciplines. Such isolationist strategies recall Marx's exposure of the wisdom of German historians in the Preface to *The German Ideology*:

> This indicates the spiritual ancestry of the great historical wisdom of the Germans who, when they run out of positive material and can serve up neither theological nor political nor literary rubbish, seize upon, not history, but rather the "prehistorical era," though without enlightening us about how one proceeds from this nonsensical "prehistory" to enter upon history proper; whilst on the other hand, their historical speculations seize especially upon this "prehistory" because here they believe themselves secure against interference by "raw facts" and so can give full rein to their speculative impulses and set up and knock down hypotheses by the thousand. (MEGA 1,5:18.5–17; CW 5:52)[6]

As Kant recognized, understanding human cognition requires identifying and understanding our basic cognitive capacities and understanding how their functioning can satisfy those normative constraints upon perception and cognition to afford—when they do—genuine cognizance, knowledge, and understanding, both commonsensical and scientific (Westphal 2004, 2021).

SOME CHARACTERISTIC PROBLEMS FOSTERED BY NEGLECTING PRACTICAL REALISM

Bas van Fraassen (2002, 18, 30) hails the astonishing "critical armamentarium" developed by analytic philosophy. Analytical philosophy has, indeed, developed indispensable critical resources, though they are neither complete nor sufficient, especially regarding issues concerning human knowledge. I provide three brief examples, indicating that for all their focus upon logical rigor, prominent analytical philosophers are prone to neglect basic logical fallacies at the core of celebrated attempts to grapple with human knowledge and its objects. These fallacies betray failures of critical self-assessment, both in logical detail and in philosophical orientation (per Griffin).

One example is Putnam's original case for his "internal realism." In "Models and Reality," Putnam (1980) examined a prevalent model-theoretic account of language and meaning. He rightly identified the fatal step in assimilating natural language to model-theoretic semantics, which requires supposing that our language can have a "full program of use" and yet lack an "interpretation," that is, a specific set of relations between terms, phrases, or sentences and their objects. He was also right that, in this model-theoretic view of language, the problem of establishing the interpretation of a language "can only have crazy solutions." However, his "internal realism," according to which truth and ontology—the way the world is—are relative to the language we use, is itself crazy. Putnam (1980) held that the meaning of our terms is given by their use, and that their use is solely a function of their (formalizable) syntax.[7] Given these assimilations, once we have understood our language, we still must "interpret" it, that is, we must still map our terms onto the world (or its features) by constructing satisfaction relations (Putnam 1976–77/1978, 494, 495; 1975–76, 188–93; 1980/1983, 19–23). This predicament shows how fully independent thoughts or concepts are from things, according to this model-theoretic approach to natural language. However, this predicament counts as a *reductio ad absurdum* of that approach to analyzing natural language in terms of logical syntax (and, we may add, logical semantics = logical explication of intension), *not* as a premiss in any sound argument for "internal realism," simply because identifying linguistic usage with formalized syntax and intension(s) is spurious. Natural languages are learned, developed, and *used* through concurrent referential, ascriptive, and descriptive uses of terms and sentences within our *actual* worldly and social context(s) (cf. Sellars 1947, 1948, 1968, 18–19).[8] To link use and reference via such "non-realist semantics" (Putnam 1980/1983, 22–23, 24) disregards, contravenes, and occludes the natural and social conditions—environmental, physiological, psychological, neonatal, and educational—requisite to any human being learning, understanding, and using language. This fallacy is fundamental, though it received no attention in the extensive critical discussion of Putnam's internal realism.[9]

A second example is Michael Friedman's celebrated *Foundations of Space-Time Theories*, in which (inter alia) he claims to demonstrate how Newton's causal theory of universal gravitation can be recast without appeal to causal force by developing suitable frameworks of "Newtonian kinematics" (Friedman 1983, 93) which can "geometrize away" gravitational forces "by incorporating the gravitational potential into the affine connection" (Friedman 1983, 95). Friedman's reformulations of Newton's mechanics retain Newton's "acceleration," only as a purely kinematic relation regarding change of velocity over time. Newton's mechanics, however, provides a comprehensive dynamic (causal) explanation of kinematic regularities throughout our solar system, including near the Earth's surface. Friedman fails to consider exactly *which* aspects of Newton's mechanics are properly modeled within his reformulations. One term used in his reformulations is "F," presumably designating "force." However, Friedman's reformulations are *merely* kinematic because his term "F" appears in his equations (34), (41), and (42), yet simply drops out of his equation (49); it is neither analyzed, explicated, reduced, nor replaced, it is merely omitted, although he acknowledges that "the spacelike vector field on the right-hand side of equation (34) is tied to the mass of bodies by equations (41) and (42)" (Friedman 1983, 119–20, cf. 123–24). Friedman's unwitting omission of "F" (in equation 49& *seq.*) preserves no more than Newtonian kinematics, it provides no more than a regularity account of "Newtonian" motions, and it fails to provide any theory of "action at a distance," because it fails to formulate, to represent, or hence to *measure* gravity as an explanatory, causal (dynamic) force (per Harper 2011, 2020, synopsis in Westphal 2014, § 2). Friedman's fallacious modeling, too, has been widely disregarded—starting with readers for the press. Empiricists chronically neglect that "$F = ma$" is neither an identity, nor a mathematical equality, nor is it Newton's equation. Newton (1726, 4–5; 1999, 406–7) expressly defines not *forces*, but three quantitative *aspects* of gravitational force; his Definitions 6–8 expressly define the *proportional quantities* of absolute, accelerative, and motive forces of gravitational attraction. These are three quantitative *aspects* of gravitational force, each of which is expressly proportional to the mass of bodies, of mutually attracting bodies, and the distance between any pair of bodies and their relative motion.

A third and for present purposes final example is Bas van Fraassen's appeal to a "logical law of weakening" as a key premise in his argument against scientific realism and for his alternative "constructive empiricism." According to van Fraassen's key argument (1980, 68–69, 90–91, 93–94, 100–101, 112, 115–16, 118, 124, 129, 143, 146, 151–52, 154–57),[10] both scientific realism and his favored constructive empiricism appeal to the same evidence base, which is empirical adequacy to describe, predict, retrodict, and (let us add) systematize our observations of natural phenomena. Because scientific realism asserts the truth of well-grounded scientific theories, whereas constructive empiricism only endorses accepting a scientific theory in view of its (sufficient) empirical adequacy, scientific realism makes a significantly stronger claim about (well-grounded) scientific theories than does constructive empiricism. However, of any two equally adequate accounts of the same evidence base, the weaker, less committal account is better justified by that common evidence base. Hence, he concludes, constructive empiricism is better justified than scientific realism. The third premise here van Fraassen calls "the law of weakening," which he claims is this logical principle:

If "A ⊃ B," then: "(A & C) ⊃ B," for any arbitrary "C."

Though stated truth-functionally, van Fraassen claims it holds for any and all logical conditionals, including (e.g.) strict implication. Yet he rightly recognizes that no *logical* conditional correctly formulates ordinary or also scientific if–then statements because these all contain, implicitly or explicitly, a ceteris paribus clause. There are two problems here. First, van Fraassen's "law of weakening" does not hold for any arbitrary "C"; it holds only for those "C" which are logically consistent with "A" and with "B." His "law of weakening" is *not* a logical principle, because it requires an implicit *semantic* presupposition regarding (at least) intension (meaning, classification), if not (also) extension (relevant instances). Much more seriously, all causal conditionals, whether commonsense or scientific, presuppose a ceteris paribus clause; consequently, van Fraassen's purported "*logical* law" of "weakening" is entirely *irrelevant* to causal explanations, whether commonsense or scientific. Van Fraassen's key argument for his constructive empiricism is predicated on a logical fallacy, which has now been neglected for over forty years![11]

The relevance of these examples to practical realism will emerge below; these brief summaries suffice to indicate that analytical philosophy too often lacks a sufficient *self-critical* armamentarium. Such critical self-assessment requires scrutinizing logical validity and soundness, of course, but it also requires critical reflection upon one's own methods, approaches, and orientation within their thematic, substantive contexts (cf. Toulmin 1949, 1958; Bird 1972), within their context among alternative methods and approaches and with regard to the kind of naturalism highlighted by Vihalemm's practical realism. These self-critical requirements also link practical realism to Kant's critical philosophy, *without* invoking Kant's transcendental idealism (Westphal 2004, 2021, 2024).

PRAGMATICS, REFERENCE, AND
PRACTICAL REALISM

There is an important instructive philosophical and historical lesson to be learned from Putnam's internal realism, which pertains both to the examples just summarized (see section "Some Characteristic Problems Fostered by Neglecting Practical Realism") and to practical realism. The very same issue about reducing language to formalized syntax was raised against Carnap (1931); these issues remain central to Carnap's *Logical Syntax of Language* (1934), and to all his semantic theories. Zilsel (1932) objected that Carnap's formalized syntax provides no more than an empty formalism and cannot distinguish between any one formalized syntactic structure which may be true (or may be used to formulate or state truths) and any other such structure which may be sheer fantasy or otherwise empty or cognitively vacuous. The parallel to Putnam's internal realism should be evident: like Carnap's logical syntax for any language (or language fragment, including, e.g., scientific theories), Putnam's internal realism reformulates our "language" solely in terms of logical syntax, within the "formal" mode of speech, thus prescinding from reference to any actual particulars. Carnap's reply to Zilsel was direct, decisive, and widely neglected to the present day. Carnap granted that his logical syntax only pertains to a formally (metalinguistically) reconstructed language but highlighted that any such logical syntax, or (put otherwise) any formally reconstructed language, requires for any *actual* use and for any actual *content* its proper complement, "descriptive semantics," which identifies those actual statements uttered by actual people within actual contexts at specified times and places, which can be accurately identified and characterized within a relevant formally reconstructed language. In this connection, Carnap (1932, 178) states directly that the required "descriptive semantics belongs to (physical) science; its sentences are (in general) synthetic, empirical." Such descriptive semantics is required, inter alia, to identify "actual protocol sentences," using historical concepts belonging to real sciences (Carnap 1932, 179–80, cf. 182). This key feature of Carnap's semantics remains widely neglected; in part because Carnap (1932) remains untranslated, and Carnap's and Zilsel's exchange is now considered merely as part of "the" protocol sentence debate. Nevertheless, Carnap's descriptive semantics is fundamental to his formal syntax and semantics throughout his career.[12] He indicates both the character and the importance of "descriptive semantics," and recognizes it belongs to *pragmatics*, that is, to the empirical study of actual language used by actual people in their specific contexts, in *Logical Syntax of Language* and ever after.[13] Carnap likewise contrasts pure (or formal) semantics to descriptive semantics in his *Introduction to Semantics* (1942, §§ 5, 24), again characterizing "descriptive semantics" in the same terms as his (1932) and (1934a). A further key indication is Carnap's (1942, § 7) clear recognition that explicating precisely any sentence or statement within the formal mode of speech using his logical syntax and pure semantics (intension) identifies truth-*conditions* of such sentences, whereas identifying the truth-*value* of any such sentence requires proper empirical inquiry. Any actual use of language to make any actual statement, including the actual uses required to identify the truth-value of any actual statement, all belong to *pragmatics* (Carnap 1942, § 4). In distinguishing, as he does, the three parts of semiotics: syntax, semantics, and pragmatics, Carnap accords entirely with Morris (1938). In this regard, Travis (1981) is quite right, and quite in accord with Carnap, that the True and the False belong entirely within the domain of the pragmatic. Actual practices and activities are required to make, investigate, assess, corroborate, improve, or disconfirm any true or false empirical statement, to which we may add: any (in)sufficiently approximate statement. This accords fully with the root and basis of pragmatism, central to pragmatic realism, of *pragma*, our actual practices, procedures, and activities, especially those of inquiry and assessment of results. This accords fully with Rein's practical realism: actual truths can only be investigated, ascertained, assessed, or (when need be) revised or replaced within actual practices of inquiry.

The central feature of recent neo-pragmatism (running from Quine and Goodman through Putnam, Rorty, van Fraassen, and Brandom) is deflating "pragmatics" to merely muddling through, whereas truth and falsehood are (purportedly) independent of the pragmatics of language and are assimilated entirely to the formal, metalinguistic mode of speech; any relation to any actual person, and hence to any actual person's context, belongs to (what remains of) pragmatics. This dismissive view of pragmatics is explicit in van Fraassen (1980, 4, 53–54, 57–58, 83, 87, 90–91, 100–101, 134–57), who expressly assigns to "pragmatics" any person- or context-dependence, including any and all human interests, while claiming that scientific theories can be stated in context-independent terms (van Fraassen 1980, 90–91).[14] Van Fraassen (1980, 89n8) cites Morris (1938), but completely misrepresents Morris's distinctions between syntax, semantics, and pragmatics, borrowing merely these three terms, while disregarding their original use by Morris and Carnap, especially concerning the pragmatics of actual linguistic behavior, especially within empirical inquiries.[15]

ZURÜCK ZU . . . CARNAP?

What would convince philosophers who pride themselves for their logical acumen, clarity, and rigor, that meaning = use = logical syntax? What would convince them that truth = satisfaction within a formal model? How would they be persuaded by faulty formalizations? How would they dismiss the "pragmatics" of actual use of language within actual contexts? The unequivocal answer is: Quine's ceaseless promotion of extensionalism.[16] Discerning scholars have recognized Quine was at cross-purposes with Carnap from the outset, substantively and methodologically.[17] Quine later noted his disagreement in two characteristic passages:

> Carnap maintains that ontological questions, and likewise questions of logical or mathematical principle, are questions not of fact but of choosing a convenient conceptual scheme or framework for science; and with this I agree only if the same be conceded for every scientific hypothesis. (Quine 1951, 72)

Ontological questions, under this [sc. Quine's own] view, are on a par with questions of natural science . . . the question [is] whether to countenance [a class of] . . . entities. This, as I have argued elsewhere, is the question whether to quantify with respect to variables which take . . . [such entities] as values. Now Carnap [1950b] has maintained that this is a question not of matters of fact but of choosing a convenient language form, a convenient conceptual scheme or framework for science. With this I agree, but only on the proviso that the same be conceded regarding scientific hypotheses generally. (Quine 1961b, 45)

In the latter passage, Quine claims to have argued for his view elsewhere; however, Quine's disagreement presupposes rather than justifies his radical semantic ascent to nothing but his version of a formalizable metalanguage. Why assimilate any and all (cognitively significant) language to a formal(izable) metalanguage? This Quine does not explain in print. His radical semantic ascent is presented though presumed rather than justified in the first of his 1934 "Lectures on Carnap" (Creath 1990, 47–67). Here Quine claims to present the context within which to understand Carnap's *Logical Syntax of Language*. However, this context is already Quine's own, and decidedly *not* Carnap's; here Quine proposes to assimilate *all* (cognitively significant) language, expressly including scientific language, to his proposed formalizable metalanguage.[18] This "semantic ascent" persists throughout Quine's views; this alone allows him to regard bound variables and pronouns as the basic, indeed the sole referential devices, eliminating names by using Russellian descriptions (Quine 1951, 67). Like so many others, Quine never noticed that, regardless of their specificity or their specification, any purportedly "definite" description may unwittingly describe many individuals, none at all, or by unwitting luck only one. Quine's favorite example, "the shortest spy," might describe congenital triplets, each of the same demure stature and the same clandestine profession, or it may become vacuous, should we ever manage to banish espionage altogether. That a descriptive phrase as such suffices to secure unique *reference* to any one actual individual is yet another dogma of empiricism (see section "*Gegenstandsbezogenheit*: Carnap, Newton, Kant and Practical Realism"); moreover, descriptive phrases require meaningful or significant *predicates*, for which Quine's extensionalism never did account credibly.

Because Quine formulates and examines bound variables and pronouns solely within his proposed formalizable metalanguage, reference is always and in principle "inscrutable," because referents can be reassigned ad libitum using proxy functions while preserving what Quine happens to call "truth," which provides no more than arbitrary assignments of (putative) individuals. About the inscrutability of reference, Quine stated:

> . . . the inscrutability of reference . . . admitted of conclusive and trivial proof by proxy functions, hence model theory . . . . (Quine 2000, 420)

Previously, however, Quine (1964, 73–74) demonstrated that such use of proxy functions involves no more than formal, metalinguistic senses of "definition" and of "interpretation" (qua randomly assigned (putative) reference to some suitably numerous domain of alleged particulars) and preserves no more than an extensional counterpart to the original "definitions" or "truths." Such shifts between such "systems" he expressly acknowledged to be "farcical," exploiting "definitional hocus pocus" which dispenses with any "literal reading" of the hijacked scientific language, by which he concludes that this is more bad news "for the notion of analyticity."

To the contrary, Quine demonstrated instead the utter empirical incompetence of his merely extensionalist metalinguistic proposals. Quine (1951, 70) sought to preserve extensionalism by appeal to Zermelo set theory, so that whatever sense might be found in "meaning" or "meaningfulness" can be preserved by appeal only to various sets of particulars. The problem for Quine's extensionalist ploy is that, in principle, it can provide *no* criteria for membership in any such set, so that one set of particulars can be affiliated with some one kind of particular (genera) or with some one feature of particulars (characteristics). *All* Quine's extensionalism can preserve is cardinality and purported designation of particulars, though *which* particulars he cannot specify! Carnap noted this limit of extensionalism in conversation with Tarski in 1940.[19] It is a well-known feature of formal model theory. Only loose talk of "logical truth"—a systematically misleading expression if ever there were—could seduce the incautious into accepting that use = logical syntax, that truth = designation within an arbitrary domain of particulars within formal model theory, or to be so lax about what is (not) preserved by formally modeling any theory—a key distinction rightly noted by Kaplan (1975, 772), though widely disregarded, for example, by Friedman (cf. section "Some Characteristic Problems Fostered by Neglecting Practical Realism"). Quine (1996, 159) claims to keep distinct his metalinguistic discussions of "meaningfulness" or "truth" and his own commonsense physicalist beliefs, but Quine's semantic ascent is so unstructured that he cannot sustain such distinctions, and whenever issues about truth, what exists, or about (purported) physical science arise, he reverts to his radically holistic, radically internalist views. He cannot have both without providing and *maintaining* one or more clear distinction(s), and also clear relation(s), between these two sets of his utterances, between which he shifts ad libitum. Only so could he disregard Carnap's (1955b) direct, cogent reply concerning meaning and synonymy in natural languages, including Carnap's reference to Naess (1953, 2005), who detailed such empirical methods and behavioral tests. In responding to Quine

([1960] 1963), Carnap (1963, 915, 917, 921) rightly notes the unclarity of Quine's reasoning. Yet clarity of reasoning is prerequisite for conclusiveness, especially in logical matters! Quine's extensionalist dogma and his indiscriminate use of semantic ascent are further examples of the genius Marx attributed to German historians (see section "Practical Realism and 'Naturalism'").

GEGENSTANDSBEZOGENHEIT: CARNAP, NEWTON, KANT, AND PRACTICAL REALISM

Recently, there has been greater attention to Carnap's earlier work and its philosophical context, yet some key features of both remain neglected.[20] Zilsel (1932) objected to Carnap's *Wissenschaftslogik* (per Carnap 1931) that Carnap's use of logical syntax to (re)construct a universal language for the sciences in principle cannot distinguish between any such candidate syntactic systems so as to identify one of these as *true*—a version of the "French novel" objection to coherence theories of truth. This same issue pertains to Carnap's *Logical Syntax of Language* (1934). Carnap (1932) *agrees* that his logical syntax is merely formal and devoid of any and all content *unless* and *until* it is coordinated with the findings of what Carnap calls "descriptive semantics," which belongs to the domain of *pragmatics* as the study of language(s) in actual use by flesh-and-blood human beings, including scientists. Carnap expressly assigns to "descriptive semantics" the task of collecting actual statements by actual scientists, especially their results. Indeed, Carnap (1934a, 259–60) stresses that logical syntax must be fit for use in connection with *actual* scientific language(s) to have any content or use whatsoever and must address actual scientific language(s) to be developed at all.[21] This diametrically opposes Quine's entire approach, which assimilates even scientific language(s) to his purely extensionalist metalinguistic logical point of view. Discussion of these issues has been preoccupied by verificationist theories of meaning, neglecting key *referential* features of Carnap's logical syntax. Carnap (1934a, 259–60) clarifies the way in which ordinary uses of language (the "inhaltliche" or "material" mode of speech) is so often "verschoben" (displaced): Expressions in ordinary language strongly connote that their terms pertain to *objects*, yet often there are no such objects. Accordingly, Carnap distinguishes *three* kinds of sentences:

> We shall distinguish three kinds of sentences:
>
> 1. Genuine Object-sentences. [These address not merely apparently but actually extralinguistic objects.] Example: "The rose is red."
> 2. Pseudo-object-sentences or sentences in the material ("inhaltlichen") mode of speech. [These merely appear to address extralinguistic objects,

e.g., the rose, but actually address the linguistic designation of this object, e.g., the word "rose."] Example: "The rose is a thing."
3. Syntactical sentences or sentences in the formal mode of speech. [These address some linguistic construction.] Example: "The word 'rose' designates a thing." (Carnap 1934b, 212; cf. 1934a, 12–13)[22]

The misleading character of the material mode of speech is that so often and so easily it involves pseudo-object-sentences:

Sentences in the material mode of speech feign connection to objects (*Objektbezogenheit*) where there is none. They easily lead to unclarities and pseudo-problems, indeed to contradictions. (Carnap 1934a, 14)[23]

Carnap clearly assigns the exposure of pseudo-object-sentences to logical syntax, and equally clearly assigns genuine object-sentences to the empirical sciences:

It is with syntactical and pseudo-object-sentences that logical analysis is concerned. Real-object-sentences fall within the domain of empirical science. (Carnap 1934c, 45)

Carnap's emphatic distinction between pseudo-object-sentences in the material mode of speech and genuine object-sentences speaking about actual objects shows that his response to philosophical pseudo-problems involves much more than a verificationist theory of meaning (intension, classification). Carnap's genuine object-sentences involve actual *reference* to *actual* particulars by *actual* speaking persons, especially scientists. Genuine object-sentences, Carnap (1934a, 241, 242) stresses, neither require nor afford translation into formalized metalinguistic form.[24] How then is actual reference to actual particulars specified, identified, or assessed? *By the empirical sciences*, using *their* distinctive disciplinary methods, procedures of inquiry, and *their* assessment of their observational and theoretical successes or inaccuracies.[25] Carnap (1934a, 244) does not reject the concepts "true" and "false"; he rightly recognizes that they are not *syntactical* concepts.[26] Carnap's (1934a, 260) logical syntax must be conducted in close connection with actual scientific disciplines and inquiries. These points are indicated concisely, yet cogently, by Carnap's (1942, § 7) formalized semantics (of intension, not reference or referents), where he notes that metalinguistic analysis—using his logical syntax and logical semantics— provides truth-*conditions* for sentence-forms; specifying or assessing the truth-*value* of any statement using that form of sentence requires actual empirical inquiries into the relevant, actual particulars. Carnap's views, including his triple distinction among key kinds of sentences, were heard and understood in London (cf. Carnap 1934c), though alas, not in Harvard Yard; in the United States, Carnap's sophisticated distinctions were reduced to no more than a generic distinction between a formal metalanguage and an object language—disregarding the distinction between (i) the language discussed by any metalanguage and (ii) whatever first-order language actually discusses physical objects (events, processes, phenomena, persons, etc.)—a neglect likely fostered by the misleading (and mischaracterized) contrast between "the" formal and "the" material modes of speech.

Quine (1961, 41–42) confessed being "impressed . . . with how baffling the problem has always been of arriving at any explicit theory of empirical confirmation of a synthetic statement." Quine's confessed bafflement reflects yet another dogma of empiricism: to expect some one, universal, adequate account of empirical confirmation fit for all empirical domains. Carnap, too, sought such accounts by recourse to confirmation theory or probability theory. Empiricists have been unable to account for the required inferences (cf. Kyburg 1984, 1988). Furthermore, empiricist accounts of confirmation can only address kinematics, that is, the identification of natural regularities; empiricist accounts of confirmation do not suffice for dynamics, that is, for causal inquiry into and measurements of forces which produce and hence explain those kinematic regularities. Newton's criteria of theoretical success are far more stringent than anything empiricism can provide, for example, insofar as his actual methods and criteria successfully distinguish between the mass and weight of orbiting bodies, and succeed at robust, precise converging measurements by three independent means of the inverse-square rate of gravitational attraction (Harper 2011). This was well understood by astronomers (Airy [1834] 1884; Ball 1902; Hartmann 1921) and by recent experts in HPS (starting with Stein [1967] and inter alia Chandrasekhar [1995], Harper [2011; 2020])—yet remains utterly misunderstood by (e.g.) van Fraassen's "constructive empiricism" (Westphal 2014).

Central to Newton's methodology is his fourth rule for natural (experimental) philosophy:

> In experimental philosophy, propositions gathered from phenomena by induction should be considered either exactly or very nearly true notwithstanding any contrary hypotheses, until yet other phenomena make such propositions either more exact or liable to exceptions. (Newton 1999, 796; 1726, 389)

Newton's rule aims to distinguish between rival theories or hypotheses by requiring that any such rival either improves demonstrably upon the accuracy of the current best theory in its domain or by demonstrating actual exceptions to that current best theory. Either achievement requires not merely supporting evidence but sufficiently *accurate* and *abundant* evidence to demonstrate either (or perhaps both) forms of improved accuracy. *Pace* Kuhn, Einstein's relativity theory better satisfies Newton's Rule 4 both in precision and in astronomical scope (Harper 2011, 378–85, 392; 2020); scientific "paradigms" simply are not "incommensurable" in the various ways Kuhn proposed (cf. Doppelt 1978; Scerri 2023). *Pace* Quine, satisfying Newton's ideals of theoretical success is no mere matter of metalinguistic "convenience"!

Crucial here is the fundamental *referential* requirement of Newton's Rule 4: to surpass an established theory or hypothesis in either regard requires the rival account not merely to be applic*able* to relevant natural phenomena; it must actually be referred to the relevant natural phenomena with sufficient *accuracy* to provide sufficient evidence of either kind of demonstrable improvement. Newton's Rule 4 requires *reference* to particulars—using what Carnap calls genuine object-sentences. Newton's and Carnap's referential (deictic) point coincides with, and is strongly supported by, what I call Kant's and Hegel's *Thesis of Singular Cognitive Reference*:

> No logically contingent, synthetic sentence form *has* a truth-value unless and until *S*omeone makes a statement using that form by which *S*/he *refers* that statement to (purportedly) relevant particulars. This requires both localizing that (or those) particular(s) within space and time (to sufficient approximation), and ascribing at least some manifest or measured characteristic(s) to it (or to them). These two requirements are mutually interdependent (*cf.* Evans 1975). This condition must be satisfied to ascertain the statement's accuracy, and to ascertain whether, how or how well it may be cognitively justified (Westphal 2021, § 26). Such assessments are constitutively *normative* (Westphal 2024).

This *Thesis* suffices to rule out all forms of experience-transcendent metaphysics (regardless of Kant's transcendental idealism). It also rules out all presumptions to hold actual language hostage to our metalinguistic preconceptions. I am delighted to have (finally) identified this referential aspect of Carnap's formal syntax and semantics, which stands independently of any specific account(s) of "meaning," intension, or classification, and hence independently of verificationist theories of meaning.

The mistakes about Carnap's pure syntax and semantics noted above are due to neglecting Carnap's (1934a, 260) clear insistence that his studies aim to work together *with* the scientific disciplines, and to renounce altogether any philosophical presumption to lord over the empirical sciences.[27] Such presumptions were relaunched by Quine in his first lecture purporting to report on Carnap's *Logical Syntax of Language*, and extending throughout his extensionalist, metalinguistic career. The semantic and syntactic errors exposed above pervade arguments against realism within the natural sciences, which require disregarding the pragmatics of actual use of language, especially by scientists in their imminently practical inquiries into natural phenomena.

SOME PRACTICAL REALIST CONCLUSIONS

The Thesis of Singular Cognitive Reference is one core feature of my reconstruction of Kant's epistemology which, to my profound delight, Rein found so attractive; it is one aspect of our strong convergence in views. I am happy to conclude by noting that this Thesis undercuts the same rivals to Rein's practical realism he identifies and criticizes and that it comports entirely with, and further corroborates, his Practical Realism. In particular, it corroborates these three of its five key theses:

3. Science as a theoretical activity is only one aspect of it (of science) as a practical activity whose main form is scientific experiment which in its turn takes place in the real world, being a purposeful and critically theory guided constructive, manipulative, material interference with nature;
4. Science as practice is also a social-historical activity which means, among other things, that scientific practice includes a normative aspect, too, and that means, in its turn, that the world as it is actually accessible to science is not free from norms either;
5. Though neither naïve nor metaphysical, it is certainly realism as it claims that what is "given" in the form of scientific practice is an aspect of the real world. (Vihalemm 2011, 48)

I am grateful indeed for this opportunity to examine these issues and specify these key reasons for our convergence!

NOTES

1. For example, he endorses my unqualifiedly realist reconstruction of Kant's epistemology (Vihalemm 2012, 203n4; 2013, 7, 11, 11n6), including my point that Kant's transcendental conditions of experience, knowledge, and inquiry need not be transcendentally ideal[ist] conditions—as Heidegger too had recognized in *Sein und Zeit*.

2. Specifically, in *Mind and the World Order*; cf. Westphal (2017). Regarding Dewey, see Sleeper (1966), Hare (1998, 2004), Shook (2000).

3. See Sleeper (1966), Will (1988, 1997), Pihlström (1996). Later, Sami Pihlström was persuaded by Putnam's claim about the entanglement of facts and values to retreat from the robust pragmatic realism of his first book, to develop a "naturalized

transcendental idealism." I find Putnam's claim both indiscriminate and misguided in stressing "values" rather than norms, especially the proper norms of inquiry and the self-discipline these require. Regarding cognitive norms, principles, and self-discipline, see Westphal (2024). Taking "values" as fundamental is an open invitation to chronic petitio principii and to Pyrrhonian skepticism (Westphal 2019).

4. There (Vihalemm 2007, 224n3) he observes, "Perhaps it should be noted more specifically that although in writings on philosophy of science their authors presumably eventually are seeking better elucidation of the question 'What is science?', it makes a difference whether this question is taken as a special theme of consideration or not."

5. Perhaps via sensations involved in perception (2021, 140, 141, 155); he states: "The relation between a sensation and a belief cannot be logical, since sensations are not beliefs or other propositional attitudes. What then is the relation? The answer is, I think, obvious: the relation is causal. Sensations cause some beliefs and in *this* sense are the basis or ground of those beliefs. But a causal explanation of a belief does not show how or why the belief is justified" (Davidson 2001, 143, cf. 151). Regarding meaning and belief, Davidson's (e.g., 2001, 148) postulation of a "radical interpreter" is yet another appeal to an omniscient "God's eye point of view"; such appeals are rejected by practical realism (cf. Vihalemm 2011, 48).

6. Translations from German sources are my own—K. R. W.

7. Meaning = use: Putnam (1976–77/1977, 127; 1980/1983, 4; 1981, ch. 2); use = formal syntax: Putnam (1981, ch. 2, and 1980/1983, 20–22, 24).

8. Similar points are made by Wittgenstein (1953); his view of meaning as use is no mere slogan; see de Queiroz (2023).

9. I highlighted this fallacy in Westphal (1997, xxiv–xxvi); its neglect is unsurprising as it appeared in a book destined never to be read by devoted formalists, though it has been taken up by pragmatic realists. In defense of robust pragmatic realism, I criticize Putnam's later "Carnapian Worlds" argument for internal realism in Westphal (2003). That argument is but another version of Quine's "proxy function" argument; see section "*Zurück zu* . . . Carnap?".

10. For a detailed examination and criticism of van Fraassen's argument, see Westphal (2014, §§ 5–7).

11. This fallacy is not remedied in any of his subsequent writings. His "law of weakening" is no mere principle of parsimony, which pertains solely to selecting between two equally accurate and comprehensive explanations of some one phenomenon. Van Fraassen simply asserts that scientists rely solely upon his conception of "empirical adequacy." This is false; Newton's mechanics succeeds in distinguishing between the mass and the weight of orbiting bodies, a distinction which no empiricist methodology can support (Harper 2011). Astronomers knew this of course (cf. Airy [1834] 1884; Ball 1902).

12. Though it is neglected, for example, in Wagner (2009), Creath (2012), and in many other studies.

13. Carnap's explicit, fundamental distinction between "formal" syntax and "formal semantics," on the one hand, and "descriptive semantics" on the other, is central to his *Logical Syntax of Language* (1934, 7, 14, 68, 107, 212) and to his *Introduction to*

Semantics (1942, §§ 5, 7). Though he does not use this designation in *Meaning and Necessity*, there the role of descriptive semantics is assigned to "the method of extension" (Carnap [1947] 1956, ch. 1) in regard to those statements which refer to "extra-linguistic fact" and are true or false factually (not logically). Carnap continues using the designation "descriptive semantics" in contemporaneous papers, such as "Meaning and Synonymy in Natural Language" (1955). Carnap expressly refers to and builds upon his *Logical Syntax of Language* both in his *Introduction to Semantics* and in *Meaning and Necessity*. The domain of "descriptive semantics" includes the "reist" or "thing language" *and* the "physical language" discussed in Carnap (1963a § II.4, cf. 868).

14. "Scientific theories can be stated in context-independent language. . . . So we do not need to stray into pragmatics . . . to interpret science. . . . Any factor which relates to the speaker or audience is a pragmatic factor; and if it furthermore pertains specifically to that particular linguistic situation, a contextual factor" (van Fraassen 1980, 90–91).

15. Brandom *talks* a lot about practices, but his inferentialist semantics cannot account for any such talk or practices (cf. Rosenkranz 2001; Redding 2015). In brief, Brandom's claim to generate semantic content (intension) using "incompatibility mirrors" formed by sentential negations *is* subject to Zilsel's (1932) key objection to merely metalinguistic formal syntax.

16. My remarks here are selective and concise; they augment my previous findings about Quine (Westphal 2015).

17. Creath (1987; 1990, 28–35; 1991); Parrini (2006/2021), Hardcastle (2006), Frost-Arnold (2011). Creath (1991, 354), and Wagner (2012) rightly note that Quine's version of "explication" is not Carnap's.

18. When discussing Carnap's views on ontology, Quine (1951, 67) again begins by developing his own context, which differs from Carnap's, as he notes but does not examine, much less: justify.

19. Carnap (pers. comm., March 6, 1940); Frost-Arnold (2013, 192, 140–41). Carnap (1963, 869) recalls a conversation with Quine in 1949 in which Quine recognized that classes are definable in terms of properties (*per Principia Mathematica*), so that later Carnap was surprised by Quine's (1961, 153–55) reversion to unqualified extensionalism. (Carnap cites Quine's first [1953] edition; the relevant pagination is unchanged in the second [1961]. Carnap habitually took detailed notes on his philosophical conversations with colleagues.)

20. Also in Westphal (1989, ch. 4), which this section augments; that first study shall be superseded by Westphal (forthcoming).

21. "Unsere These, daß Wissenschaftslogik Syntax ist, darf also nicht dahin mißverstanden werden, als könne die Aufgabe der Wissenschaftslogik losgelöst von der empirischen Wissenschaft und ohne Rücksicht auf deren empirische Ergebnisse bearbeitet werden. Allerdings ist die syntaktische Untersuchung eines schon gegebenen Systems eine rein mathematische Aufgabe; aber die Sprache der Wissenschaft liegt nicht in syntaktisch bestimmter Form vor; wer sie untersuchen will, muß daher auf den in der Fachwissenschaft praktisch angewendeten Sprachgebrauch achten und in Anlehnung an ihn erst die syntaktischen Bestimmungen aufstellen" (Carnap 1934a, 259–60).

22. Carnap's original: «Wir wollen drei Arten von Sätzen unterscheiden:

1. Echte Objektsätze. [Sie handeln nicht nur scheinbar, sondern wirklich von außersprachlichen Objekten.] Beispiel: "Die Rose ist rot."
2. Pseudo-Objektsätze oder Sätze der inhaltlichen Redeweise. [Sie handeln scheinbar von außersprachlichen Objekten, z. B. von der Rose, in Wirklichkeit aber von der sprachlichen Bezeichnung dieses Objekts, z. B. von dem Wort "Rose."] Beispiel: "Die Rose ist ein Ding."
3. Syntaktische Sätze oder Sätze der formalen Redeweise. [Sie handeln von einem Sprachgebilde.] Beispiel: "Das Wort 'Rose' ist eine Dingbezeichnung."» (Carnap 1934b, 212; cf. 1934a, 12–13)

23. "Die Sätze der inhaltlichen Redeweise täuschen Objektbezogenheit vor, wo keine vorhanden ist. Sie führen leicht zu Unklarheiten und Scheinproblemen, ja sogar zu Widerprüchen" (Carnap 1934a, 14).

24. "Es sei noch einmal daran erinnert, daß die Unterscheidung zwischen formaler und inhaltlicher Redeweise sich nicht auf die echten Objektsätze bezieht, also nicht auf die Sätze der Fachwissenschaften und auch nicht auf die fachwissenschaftlichen Sätze, die in Erörterungen der Wissenschaftslogik (oder der Philosophie) vorkommen. (Vgl. die drei Rubriken, S. 212.)" (Carnap 1934b, 242).

25. "[. . .] die Protokollsätze aufzustellen, ist Sache des beobachtenden, protokollierenden Physikers" (Carnap 1934b, 244). Uebel (1992, 122, cf. 137n49) is quite mistaken that in ca. 1932 Carnap's "conception of protocols now was wholly conventional (foreshadowed in part in his response to Zilsel [Carnap 1932, 179])." Carnap views the *forms* of protocol sentences, that is, their syntactic structures, as conventional; however, actual protocol statements or reports are *not*; scientists must use, make, assess, and if need be revise or discard such reports, *not* philosophers. Carnap adopts a *fallibilist* account of scientific knowledge, *not* a conventionalist account!

26. Carnap's logical syntax has been widely misunderstood in this regard, for example, by Oberdan (1992). In this same period, he had already mentioned "semantics," though understandably it required sustained research to develop his (Carnap 1942; [1947] 1956) semantic views.

27. Hence, I agree with Stein (1992) that Carnap was not entirely wrong after all; for further evidence and reasons for this, see Westphal (forthcoming).

REFERENCES

Airy, George Biddell. [1834] 1884. *Gravitation: An Elementary Explanation of the Principal Perturbations in the Solar System.* 2nd edition. London: Macmillan.
Ball, Robert. 1902. *Great Astronomers.* 2nd edition. London: Pitman & Sons.
Bird, Graham. 1972. *Philosophical Tasks.* London: Hutchinson's University Library; rpt.: New York: Routledge, 2021.
Burge, Tyler. 1999. "Comprehension and Interpretation." In *The Philosophy of Donald Davidson*, edited by L. E. Hahn, 229–50. Chicago, IL and LaSalle, IL: Open Court.

Carnap, Rudolf. 1931. "Die physikalische Sprache als Universalsprache der Wissenschaft." *Erkenntnis* 2: 432–65. https://doi.org/10.1007/BF02028172.

Carnap, Rudolf. 1932. "Erwiderung auf die vorstehenden Aufsätze von E. Zilsel und K. Duncker." *Erkenntnis* 3: 177–88. https://doi.org/10.1007/BF01886417.

Carnap, Rudolf. 1934a. *Die Aufgabe der Wissenschaftslogik. Einheitswissenschaft* 3. Vienna: Gerold & Co.

Carnap, Rudolf. 1934b. *Logische Syntax der Sprache.* Vienna: Springer. https://doi.org/10.1007/978-3-662-25375-5 (English translation: Carnap, Rudolf. 1959. *The Logical Syntax of Language*, translated by A. Smeaton. Paterson, NJ: Littlefield, Adams, & Co.)

Carnap, Rudolf. 1934c. "Report of Lectures on *Philosophy and Logical Syntax.* Delivered on 8, 10 and 12 October at Bedford College in the University of London, by Professor Rudolf Carnap." *Analysis* 2 (3): 42–48. https://doi.org/10.1093/analys/2.3.42.

Carnap, Rudolf. 1942. *Introduction to Semantics.* Cambridge, MA: Harvard University Press.

Carnap, Rudolf. [1947] 1956. *Meaning and Necessity.* 2nd revised edition. Chicago, IL: University of Chicago Press.

Carnap, Rudolf. 1950a. *Logical Foundations of Probability.* Chicago, IL: University of Chicago Press.

Carnap, Rudolf. [1950b]. "Empiricism, Semantics and Ontology." *Revue Internationale de Philosophie* 4; 2nd rev. ed. in: idem. (1956), 205–21.

Carnap, Rudolf. 1955. "Meaning and Synonymy in Natural Languages." *Philosophical Studies* 6 (3): 33–47. https://doi.org/10.1007/BF02330951.

Carnap, Rudolph. 1956. *Meaning and Necessity*, 2nd rev. ed. Chicago, University of Chicago Press.

Carnap, Rudolf. 1959. *The Logical Syntax of Language*, translated by A. Smeaton. Paterson, NJ: Littlefield, Adams, & Co. (English translation of Carnap, Rudolf. 1934a. *Logische Syntax der Sprache.* Vienna: Springer.)

Carnap, Rudolf. 1963. "Reply to Quine." In *The Philosophy of Rudolf Carnap*, edited by P. A. Schilpp, 914–22. LaSalle, IL: The Library of Living Philosophers.

Chandrasekhar, Subrahmanyan. 1995. *Newton's Principia for the Common Reader.* Oxford: The Clarendon Press.

Creath, Richard. 1987. "The Initial Reception of Carnap's Doctrine of Analyticity." *Nous* 21: 477–99. https://doi.org/10.2307/2215669.

Creath, Richard, ed. 1990. *The Quine-Carnap Correspondence and Related Work.* Berkeley, CA: University of California Press.

Creath, Richard. 1991. "Every Dogma Has Its Day." *Erkenntnis* 35: 347–89. https://doi.org/10.1007/BF00388294.

Creath, Richard, ed. 2012. *Rudolf Carnap and the Legacy of Logical Empiricism.* Dordrecht: Springer. https://doi.org/10.1007/978-94-007-3929-1.

Davidson, Donald. 2001. *Subjective, Intersubjective, Objective.* Oxford: The Clarendon Press.

de Queiroz, Ruy J. G. B. 2023. "From *Tractatus* to Later Writings and Back—New Implications from the *Nachlaß*." *SATS—Northern European Journal of Philosophy* 24 (2): 167–203. https://doi.org/10.1515/sats-2022-0016.

Doppelt, Gerald. 1978. "Kuhn's Epistemological Relativism: An Interpretation and Defense." *Inquiry* 21 (1–4): 33–86. https://doi.org/10.1080/00201747808601839.

Dretske, Frederick. 1981. *Knowledge and the Flow of Information.* Cambridge, MA: MIT Press.

Evans, Gareth. 1975. "Identity and Predication." *Journal of Philosophy* 72 (13): 343–63. https://doi.org/10.2307/2025212. Reprinted in Evans, Gareth. 1985. *Collected Papers*, 25–48. Oxford: The Clarendon Press.

Evans, Gareth. 1985. *Collected Papers.* Oxford: The Clarendon Press.

Frost-Arnold, Greg. 2011. "Quine's Evolution from 'Carnap's Disciple' to the Author of 'Two Dogmas.'" *HOPOS: The Journal of the International Society for the History of Philosophy of Science* 1 (2): 291–316. https://doi.org/10.1086/660011.

Frost-Arnold, Greg, ed. 2013. *Carnap, Tarski, and Quine at Harvard: Conversations on Logic, Mathematics, and Science.* Chicago, IL: Open Court.

Hardcastle, Gary. 2006. "Quine's 1934 'Lectures on Carnap.'" Draft manuscript. Accessed October 2023. https://philsci-archive.pitt.edu/2908/.

Hare, Peter H. 1998. "Classical Pragmatism, Recent Naturalistic Theories of Representation and Pragmatic Realism." In *The Role of Pragmatics in Contemporary Philosophy*, edited by P. Weingartner, G. Schurz, and G. J. W. Dorn, 58–65. Vienna: Hölder-Pichler-Tempsky.

Hare, Peter H. 2004. "Dewey, Analytic Epistemology, and Biology." In *Dewey, Pragmatism and Economic Methodology*, edited by E. L. Khalil, 144–52. London: Routledge.

Hare, Peter H., and Chandana Chakrabarti. 1980. "The Development of William James's Epistemological Realism." In *History, Religion, and Spiritual Democracy: Essays in Honor of Joseph L. Blau*, edited by M. Wohlgelernten, 231–46. New York: Columbia University Press. https://doi.org/10.7312/wohl91446-018.

Hartmann, Jul, ed. 1921. *Astronomie.* In *Die Kultur der Gegenwart*, edited by P. Hinneburg, 3. Teil, 3. Abt., Bd. 3. Leipzig and Berlin: Teubner.

Kaplan, David. 1975. "How to Russell a Frege-Church." *The Journal of Philosophy* 72 (19): 716–29. https://doi.org/10.2307/2024635.

Kyburg, Henry E., Jr. 1984. *Theory and Measurement.* Cambridge: Cambridge University Press.

Kyburg, Henry E., Jr. 1988. "The Justification of Deduction in Science." In *The Limitations of Deductivism*, edited by A. Grünbaum and W. Salmon, 61–94. Berkeley, CA: University of California Press.

Marx, Karl. 1845. "Theses on Feuerbach." In MEGA 1,5: 533–35; CW 5: 3–8.

Marx, Karl, and Friedrich Engels. 1845–46. *The German Ideology.* In MEGA 1,5: 1–528; CW 5: 19–539.

Marx, Karl, and Friedrich Engels (MEGA). 1975–. *Marx–Engels Gesamtausgabe*, 90 vols. Berlin: de Gruyter (incomplete).

Marx, Karl, and Friedrich Engels (CW). 1975–2004. *Collected Works*, 50 vols. Moscow: Progress Publishers; New York: International Publishers.

Merleau-Ponty, Maurice. 2005. *Phénoménologie de la perception.* 2nd revised edition. Paris: Gallimard.

Merleau-Ponty, Maurice. 2012. *Phenomenology of Perception*, translated by D. A. Landes. London and New York: Routledge. https://doi.org/10.4324/9780203720714.

Morris, Charles. 1938. *Foundations of the Theory of Signs*. In *Foundations of the Unity of Science: Towards an International Encyclopedia of Unified Science*, edited by Otto Neurath, Rudolph Carnap, and Charles F. W. Morris, vol. 1, 73–137. Chicago, IL: University of Chicago Press.

Naess, Arne. 1953. *Interpretation and Preciseness: A Contribution to the Theory of Communication*. Skrifter Norske Videnskaps-Akademi. Oslo, II. Hist.-Philos. Klasse, No. 1. Oslo: Norske Videnskaps-Akademi; rpt. in: idem. (2005), vol. 1.

Naess, Arne. 2005. *The Selected Works of Arne Naess*, edited by H. Glasser and A. Drengson, 10 vols. Dordrecht: Springer.

Newton, Isaac. 1726. *Philosophiae naturalis principia mathematica*. 3rd revised edition. London. Reprint 1871. Glasgow: Maclehose.

Newton, Isaac. 1999. *The Principia: Mathematical Principles of Natural Philosophy*, translated by I. Bernard Cohen and Anne Whitman, with assistance from Julia Budenz. Berkeley, CA: University of California Press.

Oberdan, Thomas. 1992. "The Concept of Truth in Carnap's *Logical Syntax of Language*." *Synthese* 93 (1–2): 239–60. https://doi.org/10.1007/BF00869427.

Parrini, Paolo. 2006. "Analiticità e olismo epistemologico: alternative praghesi." In *Le ragioni del conoscere e dell'agire. Scritti in onore di Rosaria Egidi*, edited by R. M. Calcaterra, 190–204. Milano: Franco Angeli.

Parrini, Paolo. 2021. "Analyticity and Epistemological Holism: Prague Alternatives." *Philosophical Inquiries* 9 (1): 79–94. https://doi.org/10.4454/philinq.v9i1.XXX. (English translation of previous item.)

Pihlström, Sami. 1996. *Structuring the World: The Issue of Realism and the Nature of Ontological Problems in Classical and Contemporary Pragmatism*. Acta Philosophica Fennica 59. Helsinki: Societas Philosophica Fennica.

Putnam, Hilary. 1975–76. "What Is Realism?" *Proceedings of the Aristotelian Society* 76: 177–94. https://doi.org/10.1093/aristotelian/76.1.177.

Putnam, Hilary. 1976–77. "Realism and Reason." *Proceedings and Addresses of the American Philosophical Association* 50 (6): 483–98. Reprint in Putnam, Hilary. 1978. *Meaning and the Moral Sciences*, 123–38. London: Routledge & Kegan Paul.

Putnam, Hilary. 1978. *Meaning and the Moral Sciences*. London: Routledge & Kegan Paul.

Putnam, Hilary. 1980. "Models and Reality." *The Journal of Symbolic Logic* 45 (3): 464–82. Reprint in Putnam, Hilary. 1983. *Realism and Reason: Philosophical Papers*, vol. 3, 1–25. Cambridge: Cambridge University Press. https://doi.org/10.1017/CBO9780511625275.003.

Putnam, Hilary. 1981. *Reason, Truth, and History*. Cambridge, MA: Harvard University Press. https://doi.org/10.1017/CBO9780511625398.

Putnam, Hilary. 1983. *Realism and Reason: Philosophical Papers*, vol. 3. Cambridge: Cambridge University Press. https://doi.org/10.1017/CBO9780511625275.

Quine, W. V. O. 1951. "On Carnap's Views on Ontology." *Philosophical Studies* 2 (5): 65–72. https://doi.org/10.1007/BF02199422.

Quine, W. V. O. 1960. "Carnap and Logical Truth." *Synthese* 12 (4): 350–74. https://doi.org/10.1007/BF00485423 (Reprint in Schilpp, P. A. 1963. *The Philosophy of Rudolf Carnap*, 385–406. LaSalle, IL: The Library of Living Philosophers.)

Quine, W. V. O. 1961a. *From a Logical Point of View.* 2nd revised edition. New York: Harper & Row.

Quine, W. V. O. 1961b. "On What There Is." In *From a Logical Point of View.* 2nd revised edition, 1–19. New York: Harper & Row.

Quine, W. V. O. 1961c. "Two Dogmas of Empiricism." In *From a Logical Point of View.* 2nd revised edition, 20–46. New York: Harper & Row.

Quine, W. V. O. 1964. "Implicit Definition Sustained." *Journal of Philosophy* 62 (2): 71–74. https://doi.org/10.2307/2023271.

Quine, W. V. O. 1996. "Progress on Two Fronts." *The Journal of Philosophy* 93 (4): 159–63. https://doi.org/10.2307/2940885.

Quine, W. V. O. 2000. "Response to Horwich." In *Knowledge, Language and Logic: Questions for Quine*, edited by A. Orenstein and P. Kotatko, 419–20. Dordrecht: Kluwer (Springer).

Redding, Paul. 2015. "An Hegelian Solution to a Tangle of Problems Facing Brandom's Analytic Pragmatism." *British Journal for the History of Philosophy* 33 (4): 657–80. https://doi.org/10.1080/09608788.2014.984284.

Rosenkranz, Sven. 2001. "Farewell to Objectivity: A Critique of Brandom." *The Philosophical Quarterly* 51 (203): 232–37. https://doi.org/10.1111/j.0031-8094.2001.00226.x.

Scerri, Eric. 2023. "A New Response to Wray and an Attempt to Widen the Conversation." *Substantia. An International Journal of the History of Chemistry* 7 (1): 35–43. https://doi.org/10.36253/Substantia-1806.

Schilpp, P. A., ed. 1963. *The Philosophy of Rudolf Carnap.* LaSalle, IL: The Library of Living Philosophers.

Sellars, Wilfrid. 1947. "Pure Pragmatics and Epistemology." *Philosophy of Science* 15 (3): 181–202. https://doi.org/10.1086/286945.

Sellars, Wilfrid. 1948. "Concepts as Involving Laws and as Inconceivable Without Them." *Philosophy of Science* 15 (4): 287–315. https://doi.org/10.1086/286997.

Sellars, Wilfrid. 1968. *Science and Metaphysics.* London: Routledge & Kegan Paul.

Shook, John R. 2000. *Dewey's Empirical Theory of Knowledge and Reality.* Nashville: Vanderbilt University Press.

Sleeper, Ralph. 1986. *The Necessity of Pragmatism: John Dewey's Conception of Philosophy.* New Haven, CT: Yale University Press.

Stein, Howard. 1967. "Newtonian Space-Time." *The Texas Quarterly* 10 (3). Reprint in Palter, R., ed. 1970. *The* Annus Mirabilis *of Sir Isaac Newton 1666–1966*, 258–84. Cambridge, MA: MIT Press.

Toulmin, Stephen. 1949. "A Defence of Synthetic Necessary Truth." *Mind* 58 (230): 164–77. https://doi.org/10.1093/mind/LVIII.230.164.

Toulmin, Stephen. 1958. *The Uses of Argument.* Cambridge: Cambridge University Press (Reprint with new Preface, 2003.)

Travis, Charles. 1981. *The True and the False: The Domain of the Pragmatic.* Amsterdam: Benjamins. https://doi.org/10.1075/pb.ii.2.

Van Fraassen, Bastian C. 2002. *The Empirical Stance.* New Haven, CT: Yale University Press.

Vihalemm, Rein. 1966. "O stupenyakh poznaniya." *Trudy po filosofii. Tartu Ülikooli toimetised* 10 (187): 13–27. English translation in Vihalemm, Rein. 2022. "On Stages of Cognition." Translated by A. Lazutkina. *Acta Baltica Historiae et Philosophiae Scientiarum* 10 (1): 138–55. https://doi.org/10.11590/abhps .2022.1.08.

Vihalemm, Rein. 2007. "Philosophy of Chemistry and the Image of Science." *Foundations of Science* 12 (3): 223–34. https://doi.org/10.1007/s10699-006-9105-0.

Vihalemm, Rein. 2011. "Towards a Practical Realist Philosophy of Science." *Baltic Journal of European Studies* 1 (9): 46–60.

Vihalemm, Rein. 2012. "Practical Realism: Against Standard Scientific Realism and Anti-Realism." *Studia Philosophica Estonica* 5 (2): 7–22. https://doi.org/10.12697 /spe.2012.5.2.02.

Vihalemm, Rein. 2013. "Interpreting Kant's Conception of Proper Science in Practical Realism." *Acta Baltica Historiae et Philosophiae Scientiarum* 1 (2): 5–14. https://doi.org/10.11590/abhps.2013.2.01.

Vihalemm, Rein. 2022. "On Stages of Cognition." Translated by A. Lazutkina. *Acta Baltica Historiae et Philosophiae Scientiarum* 10 (1): 138–55. https://doi.org/10 .11590/abhps.2022.1.08.

Wagner, Pierre, ed. 2009. *Carnap's Logical Syntax of Language.* Houndsmill: Palgrave Macmillan. https://doi.org/10.1057/9780230235397.

Wagner, Pierre. 2012. "Natural Languages, Formal Languages, and Explication." In *Carnap's Ideal of Explication and Naturalism*, edited by Pierre Wagner, 175–89. Houndsmill: Palgrave Macmillan. https://doi.org/10.1057/9780230379749_13.

Westphal, Kenneth R. 1989. *Hegel's Epistemological Realism: A Study of the Aim and Method of Hegel's Phenomenology of Spirit.* Philosophical Studies Series 43, edited by K. Lehrer. Dordrecht: Kluwer.

Westphal, Kenneth R. 1997. "Frederick L. Will's Pragmatic Realism: An Introduction." Critical editorial introduction to Will, Frederick L. 1997. *Pragmatism and Realism.* Foreword by Alasdair MacIntyre, edited by K. R. Westphal, xiii–lxi. Lanham, MD: Rowman & Littlefield.

Westphal, Kenneth R. 2003. "Can Pragmatic Realists Argue Transcendentally?" In *Pragmatic Naturalism and Realism*, edited by J. Shook, 151–75. Buffalo, NY: Prometheus.

Westphal, Kenneth R. 2014. "Hegel's Semantics of Singular Cognitive Reference, Newton's Methodological Rule Four and Scientific Realism Today." *Philosophical Inquiries* 2 (1): 9–65. Accessed October 2023, http://philinq.it/index.php/philinq/ article/view/86/ 44.

Westphal, Kenneth R. 2017. "Empiricism, Pragmatic Realism and the *A Priori* in *Mind and the World Order.*" In *Contemporary Perspectives on C.I. Lewis: Pragmatism in Transition*, edited by C. Sachs and P. Olen, 169–98. London: Palgrave Macmillan/Springer Nature. https://doi.org/10.1007/978-3-319-52863-2_8.

Westphal, Kenneth R. 2019. "Cosmopolitanism Without Commensurability: Why Incommensurable Values Are Worthless." *Jahrbuch für Recht und Ethik / Annual Review of Law and Ethics* 27: 243–66.

Westphal, Kenneth R. 2020. "A Snapshot from London of Philosophy *ca.* 1880." Posted on the author's website (https://ae-eu.academia.edu/KennethWestphal) under "Reference & Research Materials."

Westphal, Kenneth R. 2021. *Kant's Critical Epistemology: Why Epistemology Must Consider Judgment First.* Foreword by Paolo Parrini. New York and London: Routledge. https://doi.org/10.4324/9781003082361.

Westphal, Kenneth R. 2024. "The Question Answered: What *is* Kant's 'Critical Philosophy'?" In *The History of Philosophy as Philosophy: The Russian Vocation of Nelly V. Motroshilova,* edited by M. F. Bykova. Leiden: Brill.

Westphal, Kenneth R. Forthcoming. *Epistemological Realism: Hegel's Strictly Internal Critique of Epistemology in the 1807 Phenomenology of Spirit.* Supersedes Westphal, Kenneth R. 1989. *Hegel's Epistemological Realism: A Study of the Aim and Method of Hegel's Phenomenology of Spirit.* Philosophical Studies Series 43, edited by K. Lehrer. Dordrecht: Kluwer.

Will, Frederick L. 1988. *Beyond Deduction: Ampliative Aspects of Philosophical Reflection.* New York and London: Routledge.

Will, Frederick L. 1997. *Pragmatism and Realism.* Foreword by Alasdair MacIntyre, edited by Kenneth R. Westphal. Lanham, MD: Rowman & Littlefield.

Wittgenstein, Ludwig. 1953. *Philosophische Untersuchungen/Philosophical Investigations,* translated by G. E. M. Anscombe, P. M. S. Hacker, and J. Schulte. 4th revised edition. 2009. Chichester: Wiley-Blackwell.

Zilsel, Edgar. 1932. "Bemerkungen zur Wissenschaftslogik." *Erkenntnis* 3: 143–61. https://doi.org/10.1007/BF01886415.

Part II

METAPHYSICS OF
SCIENTIFIC PRACTICES

Chapter 4

Pluralist Realism

Where Onticity and Practice Meet

Olimpia Lombardi

In an article written with my (at the time) doctoral student and published in *Foundations of Chemistry* (Lombardi and Labarca 2005), I undertook the explicit defense of the ontic autonomy of the chemical world on the basis of a Kantian-rooted pluralist view, inspired by Hilary Putnam's internal realism (Putnam 1981).[1] From that perspective, the object of scientific knowledge is always the result of a synthesis between the categorical-conceptual frameworks embodied in scientific theories and the independent noumenal reality. However, unlike Kantian philosophy, our position admits the existence of different frameworks, both diachronically and synchronically; this leads to a pluralism that allows for the coexistence of different, even incompatible ontic domains. This view was further developed in subsequent works, particularly in its application to the relationship between chemistry and physics (Lombardi and Labarca 2006, 2011; Labarca and Lombardi 2010; Lombardi 2015).

In an article also published in *Foundations of Chemistry*, Rein Vihalemm (2011a) directed his attention to our first two works on the matter (Lombardi and Labarca 2005, 2006), taking a critical stance regarding the Kantian roots of our position. Although agreeing with Putnam's rejection of God's-eye perspective to conceive knowledge, Vihalemm claims: "My main criticism of the paper by Lombardi and Labarca is, however, that there seems to be no need to take one's stand on Putnam's internal realism in order to give up God's point of view . . . and defend the autonomy of the chemical world" (Vihalemm 2011a, 100). According to him, Kantian tradition embodies an idealistic philosophical conception that stands completely at odds with any realist view about science, in particular, with his *practical realism*, as developed in previous works (Vihalemm 2003, 2005).

Presented in these terms, the two views seem to be placed at opposing and irreconcilable positions. However, when analyzed in more detail, they are not as antithetical as initially supposed. My opinion is that, on the contrary, the discrepancy is more related to philosophical labels than to substantial content. The main purpose of this chapter is to clarify the terms of the debate, in order to show that, despite their differences, my pluralist realism and Vihalemm's practical realism tend to converge to a common perspective that offers a fruitful picture of the activity of real-life science.

WHAT IS ONTOLOGY?

According to Vihalemm, the article by Lombardi and Labarca (2005) argues "for the need to present a *metaphysical ontology* in the philosophy of chemistry" (Vihalemm 2011a, 100; my emphasis). In an explicit confrontation with this position, Vihalemm rejects "the very attempt in philosophy of chemistry to place on the agenda the question that there is a need to deal with some kind of *metaphysical-ontological* underpinning of chemistry" (Vihalemm 2011a, 101; my emphasis). By denouncing the idea of "opening of the door for *metaphysical-ontological* speculations in philosophy of chemistry" (Vihalemm 2011a, 101; my emphasis), he appeals to a realistic view ingrained in the practice of science. It is hard not to sympathize with Vihalemm's insistence on taking distance from metaphysical speculations to address scientific issues. The question here is whether my pluralist view is as engaged with metaphysical speculations as Vihalemm's reading suggests.

It is certainly true that I repeatedly used the term "ontology" in my writings about a pluralist proposal. And it is also true that, etymologically, "ontology" refers to the branch of metaphysics that studies the nature of being and the structure of reality. This is the meaning that Vihalemm has in mind when he explains: "Why I am speaking about metaphysical ontology? Simply because, as you know, the term 'ontology' is usually applied in modern philosophy to the branch of metaphysics that concerns itself with 'being-as-such' or with what kinds of things really exist" (Vihalemm 2011a, 98). Vihalemm's reading shows that I was not clear enough when introducing my position because this is certainly not the meaning with which the term "ontology" was used in my writings. In fact, I always used the term to denote what is beyond language and becomes the object of our knowledge. For example, when I talk about the ontology to which a language refers, or when I describe an ontology inhabited by individuals and properties, I am not using the term "ontology" to mean "the study of" something: the term refers to a certain "realm" described by a language and susceptible to be known by a subject.

In a personal exchange after reading the book *Los Múltiples Mundos de la Ciencia* (Lombardi and Pérez Ransanz 2012), Roberto Torretti, always very fond of the precise use of language, criticized our etymologically non-rigorous use of the term "ontology" and its cognates. I seriously took into account his remark, and since then I use the term "ontic domain" to talk about what I previously called "ontology": "ontic" dispenses with the suffix "logy" so that any relation with the idea of "the study of" disappears; "domain" tries to recover the idea of a territory, a region distinctively marked by some physical feature (see *Merriam-Webster's Dictionary*). In this way, when I ask about the ontic domain of chemistry, my question has no metaphysical involvement. My requirement is very plain and simple: I only want to understand how the realm described by chemistry looks like.

And, what about the meaning of the term "metaphysics" in the idea of "metaphysical ontology"? Vihalemm adopts the definition of the *Oxford Dictionary of Philosophy*, according to which metaphysics is the philosophical discipline that "raises questions about reality that lie beyond or behind those capable of being tackled by the methods of science" (Vihalemm 2011a, 98). This definition is not far from our characterization of "metaphysical" as what "is beyond any possible evidence," not only empirical but also "formal, historical and pragmatic evidence" (Labarca and Lombardi 2010, 156; see also Lombardi and Labarca 2011). But perhaps even more important than the definition is the assessment of the role of metaphysics in the discussion about the world described by science: in this regard, our position also follows a Kantian inspiration. Let us recall Kant's rejection of "metaphysica generalis" in the Transcendental Analytic, where he argues against any attempt to acquire knowledge of "objects in general" through the formal concepts and principles of the understanding, taken by themselves alone. It is in this context that Kant suggests that assuming the possibility of having unmediated intellectual access to objects (of having "non-sensible" knowledge) amounts to conflate "phenomena" with "noumena." On the same basis, Kant rejects the more specialized branches of metaphysics in the Transcendental Dialectic: any attempt to apply the concepts and principles of the understanding independently of the conditions of sensibility (i.e., any transcendental use of the understanding) is illicit.

It is precisely under the influence of this Kantian inspiration that my pluralist perspective is completely opposed to a metaphysical ontology and, as it will be discussed in the next section, stands in explicit confrontation with metaphysical realism. And, at the same time, far from being in conflict with Vihalemm's view, my pluralism is very close to his rejection of ontology and metaphysics.

WHICH REALISM?

In several points of his works, Vihalemm explicitly rejects Putnam's internal realism due to its debt to Kantian philosophy: "Putnam's internalist realism, which belongs to the tradition of Kantianism, is actually not realism at all as it scarcely succeeds in avoiding conceptual idealism without a rational reconstruction" (Vihalemm 2011a, 101; see also Vihalemm 2011b, 49, 54). In the eyes of Vihalemm, my pluralist view transitively inherits the sin of idealism from Kantian philosophy and Putnam's internalism, and this is a sufficient reason to discard it and search for better alternatives. What I will try to show in this section is that the divergences, stuck to different "ism" labels, are more related to different interpretations of Kantian writings than to substantial philosophical disagreements.

Let us begin with the notion of realism, which, as that of "being" in Aristotle, is said in many ways. In the history of philosophy, the term "realism" has been used with at least three different meanings: in relation to the problem of universals, as opposed to *nominalism*; in relation to the problem of the theoretical terms of science, as opposed to *instrumentalism*; and in a more general philosophical sense, referred to the existence of and epistemic access to a reality external to the subject, as opposed to *idealism*. Although sometimes the three meanings turn out to be related, in the present discussion, the focus will be on the third sense.

Nevertheless, even in the case of this specific philosophical meaning, different forms of realism can be distinguished:

- *Minimal realism* is the position that admits the existence of an external reality, which is totally independent of the subject of knowledge. Those who deny minimal realism adopt some kind of idealism, such as subjective idealism (Berkeley) or absolute idealism (Hegel).
- *Metaphysical realism*, in turn, not only accepts the existence of an external reality independent of the subject but conceives it as a "ready-made" world, that is, as a totality of fundamental items, with essential properties and relationships, and structured into absolute ontological categories and kinds.
- Finally, *epistemic realism* presupposes metaphysical realism but adds the assumption that it is possible to know, at least approximately, the "ready-made" world, which admits a single true description. In this sense, epistemic realism adopts a conception of truth as metaphysical correspondence: although truth is a relationship between language and reality, the truth value of the propositions of a language depends on the fundamental ontology.

Since metaphysical and epistemic realisms usually go hand in hand, I used to subsume both under the single heading of *metaphysical realism*. For metaphysical realism, then, reality is understood as a totality of objects, with its own identity and its intrinsic structure, which exists with complete independence from the subject of knowledge; and the epistemic enterprise consists in approaching as much as possible the true and complete description of such an independent reality.

In turn, on the side of metaphysical realism, there is a radical group, that of the "scientificists," who believe in the absolute—in the sense of transcending any local perspective—nature of scientific knowledge. As Torretti describes them: "Paradoxically, they speak persistently of the reality of the *external* world, as if they were disembodied spirits contemplating it from the outside, and, for all their godlessness, they put forward a view of it that is only conceivable from the standpoint of an omniscient God" (Torretti 2000, 114). According to this version of metaphysical realism, fundamental science is capable of "converging" toward the ultimate theory that supplies the true and complete description of the world:

> "Scientific realists" believe that reality is well-defined, once and for all, independently of human action and human thought, in a way that can be adequately articulated in human discourse. They also believe that the primary aim of science is to develop just the sort of discourse which adequately articulates reality—which, as Plato said, "cuts it at its joints"—, and that modern science is visibly approaching the fulfillment of this aim. (Torretti 2000, 114; emphasis in the original)

In the framework of this discrimination among the different forms of realism, the positions that follow the saga of Kantian philosophy find easily their place. Undoubtedly, Kant rejects metaphysical realism insofar as neither the objects of experience nor their categorical structure are independent of the subject. However, he is not an idealist like Berkeley or Hegel, since he embraces minimal realism: not only is there a reality independent of the subject, but it plays an unavoidable role, as a condition of possibility, in the constitution of the ontic domain to which our knowledge refers. It is precisely these Kantian resonances that, through the reworking of Rudolf Carnap and Willard V. O. Quine, led Putnam to formulate his internal realism.

However, the above remarks do not cancel the fact that we have all learned that Kantian philosophy is a form of idealism because, although there is something independent of the subject, the access to the subject-independent reality is always permeated by the active participation of the subject. But if the object of knowledge is always the result of a synthesis between two poles, the subject-independent reality and the framework introduced by the subject of knowledge, why insist on calling this view "idealism"? This seems to be the consequence of a kind of metaphysically realist prejudice, according to which whenever the subject participates in the constitution of knowledge, the pure essence of realism breaks down; thus, the only form of realism that deserves such a name is the metaphysical one.

Perhaps not all who conceive Kant as an idealist are victims of a realist prejudice. Maybe they support their view in a particular reading of Kantian philosophy, according to which the noumenal, independent reality, although existing, plays no role in human knowledge: the object of knowledge is not partially constituted, but is *created* by the action of the categorical action of the subject. This idealist interpretation is the traditional one, preferred by many scholars. And this is the reading that Vihalemm prefers when he rejects the idea "that objects must conform to our knowledge (in the Kantian sense), not our knowledge to objects" (Vihalemm 2013a, 365).

Nevertheless, ever since 1781, the meaning and significance of Kant's "transcendental idealism" has been a subject of controversy. This is not the place to make an exegesis of Kantian philosophy, but the fact that more "realist" readings have also been proposed cannot be ignored. For instance, Kenneth Westphal (2004) argues that transcendental idealism is not, pace Kant, required for designing a critical philosophy, which, on the contrary, can be interpreted as a kind of realism. Sami Pihlström (2012), in turn, accepts a pragmatically naturalized Kantian transcendental perspective on realism. And it cannot be forgotten that Putnam himself claimed that the best way to read Kant is as proposing for the first time what he called "internal *realism*" (Putnam 1981, 60).

I also prefer to embrace a non-idealist reading by following Torretti's detailed study of Kantian philosophy and his interpretation of the role played by noumenal reality. Although, against precritical realism, Kant insists on the unknowability of noumena, he also stresses that the idea of phenomenon involves in itself the reference to something that is not a phenomenon and necessarily takes part in the constitution of knowledge:

> The purely phenomenal character of the objects of experience does not exclude, but rather implies a transcendental reality that serves them as a basis, and that, although unknowable, is not for this less *effective* . . . phenomenal objects are not mere insubstantial ghosts, . . . the perception in which their presence is manifested reveals an *effective existence*. (Torretti 2005a, 676–77; italics added)

In fact, Kant compares our access to reality to that of a judge who compels the witnesses to reply to those questions that he himself thinks fit to propose. However, according to the Kantian view, the questions do not force the content of the responses: the independent reality must answer in the same language as that in which the questions were asked, but it can always respond negatively to those questions, making manifest its active presence.

In summary, Kant's inspiration may lead us to a stance that, although not metaphysically realist, lets the independent reality play an essential role in the constitution of the ontic domain to which our knowledge refers. Nevertheless, acknowledging a Kantian influence does not force us to agree with the whole of his doctrine, in particular with the fixed character of his system of categories.

BEYOND KANT: ONTIC PLURALISM

In the short story "The Analytical Language of John Wilkins" ("El lenguaje analítico de John Wilkins"), Jorge Luis Borges describes Wilkins's curious language and compares it with those of a Chinese encyclopedia and of the Bibliographic Institute of Brussels. From that comparison, he concludes that "it is clear that there is no classification of the Universe not being arbitrary and full of conjectures. The reason for this is very simple: we do not know what thing the universe is." Nevertheless, "[t]he impossibility of penetrating the divine pattern of the universe cannot stop us from planning human patterns, even though we are conscious they are not definitive" (Borges 1952, 103).

By contrast to Kantian philosophy, Borges tells us that it is not only that we cannot access reality independently of our categorical-conceptual frameworks, but also that there is no privileged and definitive framework. After having witnessed the great revolutions and the enormous ramification of twentieth-century science, it is not easy to continue assuming that our knowledge conforms to a single framework. Therefore, Kantian views about the constitution of the object of knowledge must be complemented by a *categorical-conceptual relativity*. According to this thesis, no concept, not even the most basic categories, needs to be included in our categorical-conceptual frameworks: there is no privileged concept of object or existence. In Putnam's words: "the phenomenon of conceptual relativity . . . turns on the fact that the logical primitives themselves, and in particular the notions of object and existence, have a multitude of different uses rather than one absolute 'meaning'" (Putnam 1987, 19).

Accepting the possibility of different categorical-conceptual frameworks leads to the thesis of *ontic pluralism*: *pace* Kant, there are different ontic domains, which are equally objective in different contexts and given certain interests and purposes. This means that the question "What objects does the world consist of?" can only be posed meaningfully within the context of a particular framework. Only when we have adopted a system of categories and concepts can we assume that certain facts and objects are there to be discovered. In other words, ontic questions only make sense from the perspective of knowledge; supposing otherwise would amount to putting the cart of metaphysics before the horse of epistemology.

It is worth emphasizing that, since the privileged viewpoint of God's-eye does not exist, there is not a single "true" ontic domain: all the domains have the same metaphysical status if all of them are constituted by equally objective descriptions. Furthermore, they are not mere "epistemologized" domains as opposed to the "real" world: when there is no metaphysically objective ontic domain, the very expression "epistemologized domain" loses any content.

When Putnam advocates for his ontic pluralism, he explicitly acknowledges the influence that other authors exerted on his philosophical views. In particular, he recalls Rudolf Carnap's works in the field of inductive logic, which led him to accept that even logical primitives may have different meanings in different contexts. In his famous article "The Methodological Character of Theoretical Concepts," when addressing the question of the admissibility of the entities postulated by scientific theories, Carnap states that: "The usual ontological questions about the 'reality' (in an alleged metaphysical sense) of numbers, classes, space-time points, bodies, minds, etc., are pseudo-questions without cognitive content" (Carnap 1956, 44–45). By contrast, questions about the reality of entities as asked and answered within science—for example, the question about the reality of the electromagnetic field—can be given a "good scientific meaning" (Carnap 1956, 45) when posed in the context of the language of a scientific theory.

Another debt recognized by Putnam is that with Willard V. O. Quine's thesis about translation and reference. On the basis of his semantic analysis and his "liberal" notion of physical object, understood as the material content of any portion of space-time, Quine stresses that physical objects could be replaced in theories by their space-time coordinates with no noticeable difference:

> This change in ontology, the abandonment of physical objects in favor of pure space-time, proves to be more than a contrived example. The elementary particles have been wavering alarmingly as physics progresses. Situations arise that curiously challenge the individuality of a particle, not only over time, but even at a single time. A field theory in which states are ascribed directly to place-times may well present a better picture, and some physicists think it does. (Quine 1981, 17)

This means that certain items can be modified "without disturbing either the structure or the empirical support of a scientific theory in the slightest" (Quine 1981, 19). This flexibility shows that "all ascription of reality must come rather from within one's theory of the world; it is incoherent otherwise" (Quine 1981, 21).

The idea of an ontic pluralism is also very clear in the historicist turn, beginning in the 1960s, from Thomas Kuhn (1962) and his claim that, after a revolution, scientists "live in different worlds" to Ian Hacking (2002) and his notion of "historical ontology." Torretti (2000), in turn, expresses his agreement with Putnam's pluralism on the basis that it is the view that better explains the actual practice of science. Although from a different perspective, Ulises Moulines (2004) also argues that philosophical ontology supervenes on the ontological commitments of empirical science, which, in turn, depend on each particular theoretical framework. Although without any ontic commitments, Hasok Chang (2012) advocates for a normative epistemic pluralism, which favors multiple systems of practice in each field of study, and also claims that arguments for realism from the success of science are fully compatible with pluralism and, indeed, conducive to it (Chang 2018). These are only some of the many authors, coming from different philosophical traditions, who have approached scientific knowledge from a pluralist perspective. This is perhaps the best manifestation of the fact that, at present, ontic pluralism cannot be conceived yet as a senseless idealism at odds with science.

ABOUT TRUTH

Up to this point I have advocated for the ontic pluralism inspired by Putnam's internal realism. However, this does not prevent us from recognizing the limitations of Putnam's proposal. One of those limitations is that derived from a conflicting characterization of the notion of truth.

Since metaphysical realism is traditionally linked to the adoption of truth as correspondence, it is usually assumed that the rejection of metaphysical realism necessarily implies giving up any correspondence view of truth in favor of some kind of coherence or pragmatist conception. As his strong criticism of the "copy theory of truth" shows, Putnam agrees with this assumption, which leads him to search for a different view: "'Truth,' in an internal view, is some sort of (idealized) rational acceptability—some sort of ideal coherence of our beliefs with each other and with our experiences as those experiences are themselves represented in our belief system—and not correspondence with mind-independent or discourse-independent 'states of affairs'" (Putnam 1981, 49–50). However, according to Putnam, truth cannot be simply identified with rational acceptability: "Truth cannot simply *be* rational acceptability for one fundamental reason; truth is supposed to be a property of a statement that cannot be lost, whereas justification can be lost" (Putnam 1981, 55). In other words, in order to avoid the collapse of truth onto justification, Putnam "idealizes" rational acceptability and identifies truth with justification in *ideal epistemic conditions*: "[T]ruth is an *idealization* of rational acceptability. We speak as if there were such things as epistemically ideal conditions, and we call a statement 'true' if it would be justified under such conditions" (Putnam 1981, 55).

This is not the place to review the many criticisms that Putnam's approach to the notion of truth received from very different perspectives. However, it is not difficult to understand the negative impact of this aspect of his proposal. In fact, under the assumption that rejecting metaphysical realism implies giving up any kind of correspondence, Putnam seems to be caught in an inescapable dilemma. The first horn of the dilemma consists of embracing a purely coherent or pragmatist notion of truth, which admits the relativity of truth attributions. Then, theories turn out to be self-justified by exclusively linguistic or "internal" reasons, and the basic intuition of realism, according to which the truth of a statement depends to some extent on what exists "outside" language, gets lost. Furthermore, this perspective falls into trouble to explain that not every theory is equally "good" and that the scientific practice of observation and experimentation plays a central role in the evaluation of scientific knowledge. The second horn of the dilemma is admitting the existence of some element "trans" conceptual scheme—such as the ideal conditions of justification—in terms of which the concept of truth is defined, providing stability to truth attributions. This alternative supplies a good antidote against radical relativism, as it allows theories to be true or false in an absolute way. But, at the same time, absolute truth attributions cannot be made compatible with the conceptual relativity resulting from internal realism.

Thus raised, this dilemma seems to have no way out. There is, however, a conceptually simple loophole, which consists of challenging the assumption on which the dilemma is based. In fact, rejecting metaphysical realism requires giving up not all forms of correspondence in the definition of truth, but only the metaphysical correspondence between language and the world "as it is in itself," that is, the "ready-made" world described from the perspective of God's eye.

According to the correspondence view, a proposition is true if it corresponds to a fact. In general, the debates about this view turn around how it can be applied to different languages and what correspondence consists in. But in the language-world relationship, the pole "world" is usually not analyzed, under the assumption that it is the domain of the independent reality. Even in the case of the semantic view, the discussions regarding Tarski's Convention T, "'p' is true if and only if P," where 'p' is the name of the proposition p of a language L and P is the translation of p into the metalanguage, usually focus on the relation between language and metalanguage, how to prove biconditionals in the metalanguage, and other formal matters, disregarding the discussion about how the fact referred to by p should be conceived. But such a fact might correspond to any domain (empirical, formal, fictional, etc.), whenever it is correctly defined. Therefore, nothing prevents p from referring to facts that are constituted in a Kantian sense: a Kantian-rooted realism can retain a form of correspondence understood as adequacy to the ontic domain constituted from a certain categorical-conceptual framework. Of course, the active role of the subject in the constitution of the ontic domain to which language refers makes the relativity of truth attributions possible; however, it does not force us to drop the intuition of truth as correspondence, central to realism, in favor of coherence or pragmatist approaches.

CATEGORICAL-CONCEPTUAL
FRAMEWORK VERSUS THEORY

As explained in the previous section, one of the limitations of Putnam's proposal derives from his conception of truth. Another limitation, closely related to that one, is that resulting from a weak characterization of what he calls "conceptual scheme" and here we preferred to designate as "categorical-conceptual framework."

By showing that a Kantian-rooted realism can integrate a correspondence notion of truth in a consistent way, it is possible to retain the realistic intuition according to which truth transcends the linguistic domain to reach what exists "beyond" our representations. However, it is worth asking how what is "beyond" language and thoughts plays, or could play, any role in our ontic commitments. In particular, the following objection might be posed: if the ontic domain were constituted by the categorical-conceptual framework of a theory, then such an ontic domain would necessarily confirm the theory's claims, and therefore scientific knowledge would be self-validating, leading to absolute relativism. Perhaps Putnam was aware of this danger: although he introduces his view as a middle ground between metaphysical realism and relativism, his best arguments are directed against metaphysical realism, whereas the relativism of "anything goes" remains on the prowl.

In order to overcome this criticism, it is not necessary, as Putnam supposes, to adopt truth attributions that remain stable over the changes of the categorical-conceptual frameworks. The objection can be faced with no obstacles if the categorical-conceptual constitution of the ontic domain, on the one hand, and the attributions of truth-value to the claims of a theory, on the other hand, are carefully distinguished. And this, in turn, requires keeping the difference between categorical-conceptual framework (or conceptual scheme) and theory completely sharp, a difference that is not clear at all in Putnam's work.

Putnam uses the term "conceptual scheme" in a somewhat ambiguous way, sometimes as equivalent to "language" or "system of sentences," sometimes as equivalent to "theory." However, independently of the conception of scientific theory one embraces, syntactic, semantic, or otherwise, language and theory are different items. A language is a system of symbols used for communication; everything can be said by a language, what is true and what is false, and even what is not susceptible to be true or false. Therefore, the idea of assigning a true value to a language makes no sense. A theory, by contrast, although expressed in some language, is susceptible to be true or false; in fact, its function is to identify, among all the statements expressible in a certain language, which must be considered true.

As it will be discussed in the next section, strictly speaking, what I call "categorical-conceptual framework" is not a linguistic item. Nevertheless, a categorical-conceptual framework is expressed by a language: it manifests the structure of the ontic domain to which all the theories formulated in that language can refer. This means that, from a conceptual viewpoint, two stages must be distinguished: first, the constitution of the structure of the ontic domain, in which the categorical-conceptual framework takes part, and second, the formulation of a theory that intends to truthfully describe the specific facts occurring in that domain. But since the constitution stage is logically prior to the descriptive stage, nothing prevents the truth-values of the theory's sentences from being established in terms of its correspondence with the facts of the categorically and conceptually constituted ontic domain.

It is worth stressing again that the ontic domain does not depend exclusively on the categorical-conceptual framework: it is not a mere creation of the mind, but arises from the synthesis between the noumenal realm and our framework. Therefore, in line with an essential element of realism, the independent reality plays an unavoidable role in the constitution of the ontic domain to which our knowledge refers, a role that is clearly manifested through the scientific practices of observation and experimentation. Taking up the Kantian metaphor, the scientist is like a judge who "interrogates" the noumenal reality from the perspective of a certain theory, in particular, from the categorical-conceptual framework that this theory presupposes; in turn, such a reality must "answer" in the same language in which the question was asked, that is, with the same system of categories and concepts that the framework imposes on it. However, the independent reality reserves its right to respond *negatively* to the received questions, with the result that the theory will be modified or rejected for undoubtedly empirical reasons. The consequences of negative empirical answers may be even deeper: the accumulation of anomalies, unfulfilled expectations, and failed predictions may also lead to modifying the framework of categories and generic concepts that the theory presupposes. In both cases, the experiential material obtained through observation and experimentation practices plays an essential role in the evaluation of scientific theories.

Precisely, it is also due to the essential role played by observation and experimentation in the acquisition of scientific knowledge that this Kantian-rooted view deserves to be conceived as a form of realism. This careful discrimination between categorical-conceptual framework and scientific theory additionally supplies a better picture of the development of science, which may proceed by changing its theories on the same ontic stage, or by modifying the stage itself by a revolutionary ontic conversion. This general claim can be illustrated by a well-known historical example. In the categorical-conceptual framework of late nineteenth-century physics, physicists were able to ask themselves about the fixed relationship between the mass and the electric charge of the electron, since the ontic domain of that time included an entity conceived as "electron" and endowed with the properties of mass and electric charge. In that case, the received answer was positive. Indeed, J. J. Thomson designed an experiment—conceivable in terms of the framework then in force—which supplied the approximate result of -1.76×10^8 Coulombs/gram. But at that time nobody would have posed a question about the curvature of space-time, since such an item was not part of the available ontic domain. Approximately during the same decades, physicists tried to detect the motion of the Earth by measuring the difference between the speed of light traveling in different directions, parallel and perpendicular to the Earth's motion, a difference predicted by the physical theories accepted at that time. However, in this case, the answer was negative, since no difference was measured. In the face of this situation, two possible strategies appeared as available. The conservative one preserved the categorical-conceptual framework and, with it, the ontic domain, and modified the theory so as to account for the new experimental result. This was the alternative adopted by Lorentz and Fitzgerald. The breakthrough strategy, by contrast, revised the categorical-conceptual framework and, consequently, reshaped the ontic domain; in this new context, a completely different theory was formulated, which explained the experimental results, but now reidentified from the new framework. This was Einstein's strategy.

Summing up, empirical evidence always plays an essential role in the modification of scientific knowledge, but not always in the same way. Sometimes, the theory is modified without touching its underlying framework; in this case, the old and the new theories can be compared since both refer to the same ontic domain, and it makes perfect sense to say that a prediction of the previous theory contradicts a prediction of the later. By contrast, in "revolutionary" processes, the new theory not only rejects some of the claims of the previous theory but also modifies, at least partially, its categorical-conceptual framework and, with it, the ontic domain to which that theory referred. It is in this sense that the new framework constitutes a "new world," a world where some of the items of the previous world no longer exist, and new items enter to inhabit the new ontic domain. In this situation, theories cannot be compared only in terms of their empirical adequacy: the replacement of the old theory by the new one is usually due to the fact that the former ceases to be effective in solving problems, especially those problems that most of the scientific community considers relevant and urgent at the time. But it is worth stressing again that both the essential role of empirical evidence in scientific knowledge and the different forms of scientific change can only be understood on the basis of the careful distinction between categorical-conceptual framework and theory, a distinction that is far from clear in Putnam's internal proposal.

ON CATEGORIES AND CONCEPTS

Up to this point it is quite clear that the Kantian-rooted pluralist, advocated for here, heavily relies on the notion of a *categorical-conceptual framework*, a notion clearly related to what several philosophers, including Putnam, call a "conceptual scheme." However, the move of adding the term "categorical" to the notion is not a mere terminological whim but carries very relevant consequences.

A categorical-conceptual framework is a system of *categories* and *concepts* that, in synthesis with the noumenal reality, constitutes an ontic domain as something essentially new, in which the original components can no longer be disjoined. In Kantian terms, a categorical-conceptual framework is a condition of the possibility of knowledge, and although it is expressed through language, it is not a linguistic entity itself: the same framework can be expressed by different languages. A categorical-conceptual framework cannot be identified with the mental structure of individual subjects, since it is a system of categories and concepts shared by a community: it is shaped and stabilized through social practices, not only linguistic but also material practices of manipulation and transformation, which suppose values, interests, and common objectives. In turn, nothing prevents the same person from adopting different frameworks in different situations, being aware of the differences among those frameworks as well as between their corresponding ontic domains.

An essential step to understanding the notion of a categorical-conceptual framework is distinguishing between categories and concepts. According to Aristotle, there are ten categories: substance and nine types of properties. There are twelve Kantian categories, and they are more abstract than those of Aristotle. Independently of the differences regarding the notion of category in these authors, it is quite clear that categories are not class concepts, such as "dog" or "blue," which apply to previously identified objects; categories are not *taxa*, such as "cat," "feline," and "mammal," which classify preexisting individuals. Categories are the most basic structuring elements of both the ontic and linguistic realms, logically prior to any ordering or classification, as well as to any claim about facts and truth (see Lewowicz 2005). As a consequence, a system of categories does not establish a mere division into classes of items that are "out there," waiting to be classified. It provides a first identification of the items that populate the ontic domain, to the extent that it introduces the ontic categories to which such items belong. For example, the system of categories will tell us if the domain is inhabited by individuals, properties, and relations, or if there are no individuals *stricto sensu* but only bundles of properties. It will tell us if possibility is an ontically irreducible feature of reality or if it can be reduced to actuality. On the basis of the categories of the framework, we will be able to say if there are causal links in the domain, as well as if the ontic items can be categorized as one or multiple, and if the events are temporally arranged according to past, present, and future.

Although the categorical-conceptual framework is not a linguistic item, the difference between categories and concepts finds its counterpart in language. In fact, general terms express concepts, not only of classes or properties, but also of relations. Categories, by contrast, are manifested by the structure of language itself. Following Ludwig Wittgenstein's distinction between *saying* and *showing*, whereas language *says* something about reality through its terms, it *shows* the categories that inform and organize the ontic domain referred to by it by means of its own structure. Categories are *said* neither by nouns or predicates nor by any other type of word: "What can be shown cannot be said" (Wittgenstein 1921, Proposition 4.1212). For instance, the ontic categories of object, property, relation, fact, causation, and quantity cannot be "said" by the terms of the language but are manifested by the linguistic categories of noun, monadic predicate, *n*-adic predicate, proposition, causal connectors, and grammatical number, respectively.

Besides categories, which introduce the most basic structure of the ontic domain, the framework can also include certain very generic concepts that refer to items whose existence and/or features cannot be denied. For example, in the framework of thermodynamics, the concept of heat is essential in the sense that thermodynamics is precisely the branch of physics that deals with the behavior of heat and temperature and their relations to energy, work, radiation, and certain properties of matter. Analogously, Newtonian mechanics cannot dispense with the concept of force, since it is an essential element in Newtonian laws.

Not all the elements of a categorical-conceptual framework inhabit the same level of the system: categories and some generic concepts are placed at the most basic level. Two frameworks may well share part of their basic elements, as usually happens in the history of science. By contrast, historical examples of successive frameworks that radically differ in all their categories and concepts could hardly be found. Nevertheless, the fact that two frameworks partially agree does not preclude the possibility of a breakdown between the corresponding ontic domains due to the disagreements in the rest of the frameworks' structures. A simple case of this situation is that of theories that, although empirically equivalent, are still incompatible because their frameworks diverge at the non-observable level.

On the other hand, since categories are the most basic elements of the frameworks, they tend to be endowed with the most generality and stability through the historical development of scientific knowledge. For example, although the category of an individual object has gone into crisis in certain physical theories, it was one of the most entrenched and extended across the entire history of science. Moreover, it is a fact that the historical-social context can favor—or hinder—the incorporation of very general categories and concepts in most of the frameworks, scientific or otherwise, of a certain era. For example, the concept of evolution, practically ignored before the nineteenth century, developed into a fundamental element of the categorical-conceptual frameworks of many scientific disciplines from the second half of that century, as different as biological Darwinism and macroscopic thermodynamics. Another interesting case is the concept of probability, which, virtually absent from the picture of reality until well into modern times, became unavoidable in practically all areas of knowledge also toward the mid-nineteenth century, expanding its scope with its relation to other notions such as those of possibility, indeterminism, or indetermination. These stable categories and generic concepts play the role of the "pragmatic conception of *a priori*" proposed by Clarence Irving Lewis (1923), later retaken by Thomas Kuhn (1993), and are also related to the "historical *a priori*" of Michel Foucault (1969), which crosses the boundaries of different scientific disciplines to put limitations on what science can say about at a given historical time. They also have also interest affinities with Hacking's idea of "historical ontology" as the result of the "style of reasoning" prevailing at a certain historical time: "although styles may evolve or be abandoned, they are curiously immune to anything akin to refutation" (Hacking 2002, 192), because they introduce "new ways of being a candidate for truth or falsehood" (Hacking 2002, 189).

GENUINE INCOMMENSURABILITY

When the philosophical notion of category is recovered to explain the role played by a categorical-conceptual framework in the constitution of an ontic domain, the possibility of a deep discontinuity between different domains can be understood. Such a discontinuity implies an incommensurability that is not confined to a rearrangement of preexisting items. Incommensurability does not mean that the domains constituted by different frameworks are merely different "worlds of classes": they might even disagree in their most basic structures since they are populated by categorically different items. And it is precisely this strong form of incommensurability that sustains a genuine ontic pluralism, which rejects the metaphysical commitment to a unique reality inhabited by ultimate components interrelated according to the only real structure, to which the many "epistemologized" domains referred to by science will converge or else will finally be reduced.

The relevant difference between categories and concepts has not been sufficiently taken into account by contemporary philosophers. For instance, when reinterpreting his own notion of scientific revolution, Kuhn (1983, 1993) conceived his "*taxonomic* categories" as a condition of possibility of knowledge; however, he described them as introducing different *classifications* that block inter-paradigm translation. Putnam (1981) himself not only relies on the notion of "*conceptual* scheme," but also uses it to talk about the constitutive role of "concepts." This assimilation between categories and concepts of classes or *taxa* in philosophers of the stature of Kuhn and Putnam makes it unsurprising that Hacking (1983, 1993) interpreted their proposals in a nominalist key. Hacking stresses that nominalism is not a thesis about *existence*, but a thesis about *classification*: "It says that only our modes of thinking make us sort grass from straw, flesh from foliage. The world does not have to be sorted that way; it does not come wrapped up in 'natural kinds'" (Hacking 1983, 108). According to the nominalist, there is real stuff, independent of the mind; "he denies only that it is naturally and intrinsically sorted in any particular way, independent of how we think about it" (Hacking 1983, 108). In other words, there is a reality composed of certain basic items whose existence is independent of the subject of knowledge; the conceptual scheme only introduces different classifications of that independent reality. On this basis, Hacking considers that Putnam is a "transcendental *nominalist*" (Hacking 1983, 109; for a nominalist reading of Kuhn, see Hacking 1983, 109–11).

Here I am not interested in discussing the extent to which this nominalist reading of Putnam's internalism is correct. I tend to believe that nominalism is far from Putnam's original aim; however, his loose characterization of the notion of conceptual scheme and the lack of a clear distinction between categories and concepts justify Hacking's reading. What I do want to stress is that a nominalist interpretation of ontic pluralism, although friendly to metaphysical realists, is not interesting at all because it does not match with what happens in real science. When space and time of nonrelativistic theories are replaced by space-time, the reconfiguration of the ontic domain cannot be conceived as a mere reclassification of preexisting items (see, e.g., Sklar 1974; Earman 1989). The fact that standard quantum mechanics lacks the philosophical category of an individual, at least in its traditional sense, can hardly be explained in terms of a different way of grouping individual objects in classes (see, e.g., French 1998; da Costa and Lombardi 2014; Lombardi and Dieks 2016). The deep breakdown between structural chemistry and quantum mechanics, which frustrates any description of the concept of molecular structure in quantum terms, stands also far beyond any difference in the way in which independent items are sorted into natural kinds (see, e.g., Primas 1983; Hendry 2010; Martínez González, Fortin, and Lombardi 2019).

In the context of a detailed discussion about realism in science, Ilkka Niiniluoto (1999) presented his proposal of a "critical scientific realism" as an overcoming of Putnam's internal realism. According to Niiniluoto, a reasonable scientific realist accepts the minimal ontological assumption that there is a real world, "THE WORLD," which is independent of human minds, concepts, beliefs, and interests. THE WORLD must be distinguished from the *world-versions* related to languages: each interpreted language, or conceptual system L, whose terms acquire meaning through social conventions, determines a W_L structure consisting of objects, properties, and relationships. With a rather bizarre distinction between UFOs (*unidentified flying objects*) and IFOs (*identified flying objects*), Niiniluoto explains that:

> THE WORLD contains UFOs, which are not our constructions, or produced by us in any causal sense. But these UFOs are not "self-identifying objects" in the bad metaphysical sense feared by Putnam: they are potentially identifiable by us, as extended elements or "chunks" of the world flux, by means of continuity, similarity, and mind-independent qualities. IFOs, on the other hand, are in a sense human-made constructions, objects under a description, and hence exist only relative to conceptual schemes. (Niiniluoto 1999, 221)

In other words, THE WORLD consists of self-subsistent objects, which cannot be exhaustively known due to their infinite properties and relations. Each world-version offers a partial account of that inexhaustible independent reality:

> A UFO is not an unknowable noumenon: it is not inaccessible but rather inexhaustible, something that can be described and identified in an unlimited number of ways. An IFO is not a phenomenal veil which hides a UFO from us, and it is not the content of our knowledge about the reality. Rather, it is part of THE WORLD as described relative to a conceptual framework. (Niiniluoto 1999, 222)

It is for this reason that Niiniluoto gives up the possibility of a complete and true theory of THE WORLD and emphasizes that we should be satisfied with our world-versions, always partial and relative.

Niiniluoto's proposal was well received by those philosophers—Vihalemm among them—who were concerned with defending a pluralist realism
against the threat of Kantian idealism. However, when considered in detail,
his critical scientific realism is an elaborated version of metaphysical realism,
according to which the object of knowledge is the independent reality as it is
in itself (THE WORLD) and science may asymptotically approach the true
description of that reality, independent of the knowing subject: "It is essential
to this realist view that THE WORLD has contained and contains 'full' or
'thick' objects with all of their mind-independent features. . . . Dinosaurs did
have such properties and parts as weight, length, color, bones, skin, legs, and
eyes" (Niiniluoto 1999, 220). "Dinosaurs existed as UFOs at 100 million BC,
but the related IFOs exist only after the invention of the concept 'dinosaur'"
(Niiniluoto 1999, 221). In other words, all the world-versions contribute to
the knowledge of THE WORLD as it is in itself, supplying different perspectives on the same independent reality. But this is precisely what science,
as effectively practiced, does not show: the "world-versions" supplied by
a single scientific discipline along its history and by different scientific disciplines at the same historical time are usually so different that cannot be
conceived as versions of the same single domain. On the contrary, in many
cases, those "versions" are incompatible and, therefore, cannot be conceived
as partial descriptions of a unified realm. Science, in its everyday operation, is
not involved in discovering more and more properties of dinosaurs as UFOs,
but in designing different successful ways to interplay with what is beyond
us and escapes from our control.

Summing up, both Hacking's nominalist interpretation of internal
realism and Niiniluoto's critical scientific realism remain far from any
Kantian inspiration. According to the pluralist realism I advocate for,
there is no independent reality which our knowledge would approach: our
categorical-conceptual frameworks do not merely allow us to describe an
independent world in a certain way, but rather play an unavoidable role
in the constitution of our ontic domains, with their items and structures.
To the extent that those frameworks are not perfectly translatable into
one another, the corresponding domains are genuinely incommensurable.
This does not mean that different frameworks have nothing in common:
as already stressed, complete difference is more an exception than a rule.
Neither does this mean that scientists working in the context of different frameworks cannot communicate with each other: they can rationally
and fruitfully discuss, but they cannot overcome their differences only by
empirical means, since their theories refer to different ontic domains. It is
precisely at this point that a widely articulated concept of scientific practice needs to enter the stage.

ON PRACTICE

In recent decades, a wide philosophical movement has tended to emphasize the relevance of practice in science. Against the traditional representational view, at present many authors stress the idea of scientific knowledge as an epistemic tool, that is, as a means for intended applications. From this perspective, scientific knowledge must always be understood as actively constructed by humans through their practices.

In this relatively new philosophical context, it may seem that talking about categorical-conceptual frameworks keeps us chained to an analytical and representational style that ignores the effective practice of science. Nevertheless, it is not clear why this should be the case. In fact, when the pluralist admits the possible coexistence of different frameworks, such a claim does not mean that all possible frameworks are equivalent in real-life science. On the contrary, along history, and even within a single historical time, some categorical-conceptual frameworks become constitutive of certain scientific disciplines or fields, whereas others pass away or never come to be. But, since they are not susceptible to be true or false, this fact cannot be explained by appealing to truth-values. It is precisely here where the practice of science becomes essential: certain categorical-conceptual frameworks consolidate in the actual scientific activity due to their practical success. At this point, the widely appealed to but poorly characterized notion of "scientific practice" becomes central: the problem is to understand what is meant by "scientific practice."

Many authors find in Hacking's defense of "intervening" one of the first manifestations of the "pragmatist turn" in the philosophy of science. According to Hacking, the realist/antirealist debates at the level of representation are always inconclusive: "I suspect there can be no final argument for or against realism at the level of representation. When we turn from representation to intervention . . . anti-realism has less of a grip" (Hacking 1983, 31). Here intervention is explicitly understood in terms of "experimentation and manipulation" (Hacking 1983, xii): "So far as I'm concerned, if you can spray them, they are real" (Hacking 1983, 23). The impact of Hacking's work promoted a view that places the clue of practice in experimentation, that is, in the work of experimental scientists in their labs by means of their apparatuses and instruments. However, experiment, as a procedure where the variables of a target system can be manipulated and controlled, is not the only way in which scientists obtain empirical evidence. By contrast, in certain fields manipulation is completely impossible, so empirical evidence is obtained by observation: for instance, astronomers cannot "spray" the entities which they talk about and in whose existence they believe. This does not mean that observation is a passive enterprise: agents are actively involved in observation practices, not only through the construction of the necessary instruments, but also through the design of the observational setups. Both experimentation and observation are active practices, but speaking about *empirical practice* instead of experimental practice is more accurate when describing the actual practice in laboratories since it includes both kinds of activities.

Nevertheless, constraining the interest on empirical practice leaves out of focus another essential part of the practice of science: *theoretical practice*. In fact, many scientists carry out their entire scientific life without ever entering a laboratory or touching a measuring apparatus. Let us think, for instance, of cosmologists or quantum gravity theorists. Those theoretical scientists, however, are also involved in practices that define their own activities. Those practices are related to different aspects of the theoretical work, such as formal techniques (e.g., the use of certain mathematical tools or diagrammatic strategies), general assumptions considered as non-revisable (e.g., the conservation of energy in certain fields of physics), the paradigmatic models, by means of which a theory is taught and applied (Kuhn's "exemplars"), among many others. Of course, the theoretical results obtained by means of this kind of practices should eventually be empirically tested to acquire factual support. However, the contact between theory and empirical evidence can be very complex and indirect. For example, the existence of the Higgs boson was predicted by Peter Higgs in 1964 on the basis of symmetry considerations in the context of the formalism underlying the Standard Model of particle physics; the empirical confirmation of that existence needed to wait almost fifty years, until 2012, when scientists agreed that the data collected at the Large Hadron Collider of CERN could be interpreted as evidence for the detection of the Higgs boson. Although the conceptual link between the theoretical proposal and the experimental detection cannot be denied, the practices involved in the two stages of the history can be unified in a single system only by very indirect and weak links.

At present, scientific practice tends to blur the boundaries between theoretical and empirical activities. This happens particularly due to the central role that computer simulations play in almost all areas of science: the theoretical work needed to design models and the empirical results so obtained converge in simulation practices. For instance, even a highly theoretical field such as the study of highly unstable (chaotic) systems was strongly boosted by computer technology: although the concept of high instability was introduced by Poincaré at the end of the nineteenth century, only the advent of computers with sufficient power, capacity, and computing speed enabled scientists to overcome the limitations imposed by the analytically insolvable character of the dynamical equations representing chaotic systems. There are even specific fields whose activities are mainly defined in terms of this new empirical-theoretical activity. This is the case of the practices in quantum chemistry, where the use of simulations becomes a novel form of experimentation: in the new laboratories, the tools are computers where chemists, physicists, and mathematicians work together with the purpose of further developing the discipline (Gavroglu and Simões 2015).

When science is conceived as a human activity, scientific practice goes far beyond empirical and theoretical work. Currently, a great amount of the scientists' time is devoted to the design and evaluation of research projects with the purpose of getting financial resources for their activities. This *financial support practice* is not a mere side or accessory scientific task, but has become a central endeavor in real science, which strongly shapes the goal and content of research. This is due to the fact that, in general, scientists must justify their resource requirements in function of goals whose values need to be argued for, since they will be assessed beyond the boundaries of the scientific community. The history of the Texas Superconducting Super Collider is a good example of the interplay between goals and financial constraints: it was designed during the 1980s and the project construction began in the early 1990s, with the purpose of largely overcoming the capability of CERN's LHC. However, since the first stages of the design, a heated debate ensued about the high costs of the project. Not only did the general public begin to see the whole idea with suspicion, demanding to know what their tax money would be used for. Also, many scientists pointed out that basic research in other areas, such as materials science, was underfunded compared to high energy physics, despite the fact that those fields were more likely to produce applications with technological and economic benefits. As a result of a discussion that reached the United States Congress, the project was aborted in 1993. This example shows that the practices involved in the financial support of scientific activities cannot be ignored in a realistic picture of science as a human activity.

Through these financial-support practices, science is strongly entangled with extra-scientific goals and interests that carve out the activities of agents and institutions. For instance, although computer technology exerted a decisive influence in the "rediscovery" of instability during the 1960s, a relevant socioeconomic factor also promoted the study of highly unstable systems. During those days, in the United States certain groups with interests in agricultural and cattle-breeding acquired political power and needed a better understanding of meteorological phenomena to improve weather forecasts. But, since meteorological systems are usually highly unstable, this promoted the theoretical study of instability. It was then not a coincidence that it was a meteorologist, Edward Lorenz, who obtained, in 1963, the first quantitative results from a system of differential equations corresponding to a chaotic evolution (see Lombardi 2011).

As characterized in the Mission Statement of the Society for Philosophy of Science in Practice (SPSP): "Practice consists of organized or regulated activities aimed at the achievement of certain goals" (SPSP 2019). Thus, a system of scientific practices is a complex net, with blurred boundaries, of interrelated activities developed by the different agents involved in the social phenomenon called "science." Precisely those practices are what play an essential role in deciding what is successful in science and, with it, in fixing certain categorical-conceptual frameworks over others. Of course, such practical success does not make those frameworks true, since categorical-conceptual frameworks, like Hacking's styles of reasoning, are not something amenable to having truth-value: they are the frameworks where truth and falsity can be established by empirical means.

Up to this point, it seems to be clear that once metaphysical realism has been given up, there is no argument to oppose realism and practice. As Sami Pihlström claims: "pragmatism, as such, is no enemy of (moderate) scientific realism" (Pihlström 2008, 59). But Pihlström takes a step further when he characterizes that kind of realism in the context of a pragmatist reading of Kant's philosophy; according to him, the contemporary scientific realist should be prepared to accept a pragmatically naturalized Kantian transcendental perspective on realism. From this "naturalized" Kantism, it is practice that supplies the conditions of possibility of knowledge: "it would seem—at least this rearticulation should be available to the 'Kantian pragmatist'—that practices provide transcendental (contextual) conditions for the possibility of there being scientifically representable objects at all—for us" (Pihlström 2012, 88–89). It is interesting to notice that this recovering of Kantian philosophy appears also in some authors being strongly committed to the movement of philosophy of science in practice. For instance, Chang (2009) introduces the idea of ontological principles, which, analogously to Kant's categories, do not supply empirical knowledge but allow for the intelligibility of epistemic activities—for example, the epistemic activity of counting involves the ontological principle of discreteness. Even more recently, Mieke Boon (2017) takes Kant's idea of the condition of possibility of knowledge as a starting point for an epistemology of scientific practices, especially those which aim at knowledge for practical uses.

In summary, we have no good reasons to assume that the Kantian-rooted ideas about the constitution of our knowledge are irreconcilable with science as a practical activity. In fact, in my book in collaboration with Ana Rosa Pérez Ransanz (Lombardi and Pérez Ransanz 2012), we also appealed to the second thesis of Marx's *Theses on Feuerbach*, a quote repeatedly included by Vihalemm (2011b, 2012, 2013b) in his own works:

> The question whether objective truth can be attributed to human thinking is not a question of theory but is a practical question. Man must prove the truth—i.e., the

reality and power, the this-sidedness of his thinking in practice. The dispute over the reality or non-reality of thinking that is isolated from practice is a purely *scholastic* question. (Marx 1845, Thesis II in *Theses on Feuerbach*)

FINAL REMARKS

I began this chapter by recounting the criticisms that Vihalemm raised against the Kantian-rooted pluralist realism that we adopted in order to defend the ontic autonomy of the chemical domain. In those criticisms, he opposed our position to his practical realism, according to which reality is not accessible independently of scientific practices:

> To speak about the world outside practice means to speak about something indefinable or illusory. It is only through practice that the objective world can really exist for humans. Therefore, knowledge must be regarded as the process of understanding how the world becomes defined in practice. (Vihalemm 2012, 10)

From this departing point, I analyzed different aspects of my proposal, now on the basis of the experience gained over more than ten years of discussing and rethinking it. I stressed that talking about ontic domains of science does not commit us with a metaphysical realism that aims at accessing reality from God's eye: the object of scientific knowledge is always the result of a synthesis between the categorical-conceptual frameworks embodied in scientific theories and the independent noumenal reality. But, at the same time, the Kantian inspiration can be acknowledged from a realist viewpoint, which lets the independent reality play an essential role in the constitution of the ontic domain to which our knowledge refers. However, this Kantian inspiration does not prevent us from distancing ourselves from Kant's philosophy by accepting the possibility of different categorical-conceptual frameworks, which leads to the thesis of ontic pluralism: different ontic domains may exist, which are equally objective in different contexts and in function of certain interests and purposes. In turn, the clear distinction between categorical-conceptual frameworks and theory allows us to clarify several issues. On the one hand, it allows us to retain a notion of truth as correspondence with a constituted ontic domain and, with this, to assign to empirical evidence a positive role in the building of scientific knowledge. On the other hand, it explains why, by contrast with theories, frameworks are not amenable to be true or false; they are the scenario where truth or falsity can be assigned; they consolidate in actual science due to their success in generating knowledge. But since categorical-conceptual frameworks cannot be assessed by empirical adequacy, the practice of science enters the scene by supplying the context in which the idea of success acquires substantial content. Therefore, this pluralist realism integrates a complex and articulated notion of scientific practice, which goes beyond empirical and theoretical activities, to reach the many different and complex links between the scientists' work and their social environment in its different manifestations. Once all these issues are sufficiently clarified, it is not difficult to see that my philosophical position is not at all far from Vihalemm's practical realism. On the contrary, both views converge in essential points, and supply a non-naïve picture of the social phenomenon we call "science."

NOTE

1. I am grateful to the University of Tartu, and especially Endla Lõhkivi, for the opportunity to present this chapter.

REFERENCES

Boon, Mieke. 2017. "Philosophy of Science in Practice: A Proposal for Epistemological Constructivism." In *Logic, Methodology and Philosophy of Science: Proceedings of the fifteenth International Congress (CLMPS 2015)*, edited by Hannes Leitgeb, Ilkka Niiniluoto, Päivi Seppälä, and Elliott Sober, 289–310. London: College Publications.

Borges, Jorge Luis. 1952. "El idioma analítico de John Wilkins." In *Otras Inquisiciones*. Buenos Aires: Editorial Sur. Translated by Ruth L. C. Simms in Borges, Jorge Luis. 1971. *Other Inquisitions. 1937–1952*. Austin, TX: University of Texas Press.

Carnap, Rudolf. 1956. "The Methodological Character of Theoretical Concepts." In *Minnesota Studies in the Philosophy of Science, Vol. I: The Foundations of Science and the Concepts of Psychology and Psychoanalysis*, edited by Herbert Feigl and Michael Scriven, 38–76. Minneapolis, MN: University of Minnesota Press.

Chang, Hasok. 2009. "Ontological Principles and the Intelligibility of Epistemic Activities." In *Scientific Understanding: Philosophical Perspectives*, edited by Henk W. De Regt, Sabina Leonelli, and Kai Eigner, 64–82. Pittsburgh: Pittsburgh University Press.

Chang, Hasok. 2012. *Is Water H₂O? Evidence, Realism and Pluralism*. Boston Studies in the Philosophy of Science. Dordrecht: Springer.

Chang, Hasok. 2018. "Is Pluralism Compatible with Scientific Realism?" In *The Routledge Handbook of Scientific Realism*, edited by Juha Saatsi, 176–86. Abingdon: Routledge. https://doi.org/10.4324/9780203712498-15.

da Costa, Newton, and Olimpia Lombardi. 2014. "Quantum Mechanics: Ontology Without Individuals." *Foundations of Physics* 44: 1246–57. https://doi.org/10.1007/s10701-014-97931.

Earman, John. 1989. *World Enough and Space-Time: Absolute vs. Relational Theories of Space and Time*. Cambridge, MA: The MIT Press.

Foucault, Michel. 1969. *L'Archéologie du Savoir*. Paris: Gallimard. Translated into English by A. M. Sheridan Smith in Foucault, Michel. 1971. *The Archaeology of Knowledge & the Discourse on Language*. London: Tavistock.

French, Steven. 1998. "On the Withering Away of Physical Objects." In *Interpreting Bodies: Classical and Quantum Objects in Modern Physics*, edited by Elena Castellani, 93–113. Princeton, NJ: Princeton University Press. https://doi.org/10.1515/9780691222042-009.

Gavroglu, Kostas, and Ana Simões. 2015. "Philosophical Issues in (Sub)Disciplinary Contexts: The Case of Quantum Chemistry." In *Essays in the Philosophy of Chemistry*, edited by Eric Scerri and Grant Fisher, 60–81. Oxford: Oxford University Press. https://doi.org/10.1093/oso/9780190494599.003.0008.

Hacking, Ian. 1983. *Representing and Intervening*. Cambridge: Cambridge University Press. https://doi.org/10.1017/CBO9780511814563.

Hacking, Ian. 1993. "Working in a New World: The Taxonomic Solution." In *World Changes: Thomas Kuhn and the Nature of Science*, edited by Paul Horwich, 275–310. Cambridge, MA: MIT Press.

Hacking, Ian. 2002. *Historical Ontology*. Cambridge, MA: Harvard University Press.

Hendry, Robin Findlay. 2010. "Ontological Reduction and Molecular Structure." *Studies in History and Philosophy of Modern Physics* 41 (2): 183–91. https://doi.org/10.1016/j.shpsb.2010.03.005.

Kuhn, Thomas S. 1962. *The Structure of Scientific Revolutions*. Chicago, IL: The University of Chicago Press.

Kuhn, Thomas. 1983. "Commensurability, Communicability, Comparability." In *PSA: Proceedings of the Biennial Meeting of the Philosophy of Science Association 1982*, vol. 2, edited by Peter D. Asquith and Thomas Nickles, 669–88. East Lansing: Philosophy of Science Association. https://doi.org/10.1086/psaprocbienmeetp.1982.2.192452.

Kuhn, Thomas S. 1993. "Afterwords." In *World Changes: Thomas Kuhn and the Nature of Science*, edited by Paul Horwich, 311–41. Cambridge, MA: MIT Press.

Labarca, Martín, and Olimipa Lombardi. 2010. "Why Orbitals Do Not Exist?" *Foundations of Chemistry* 12 (2): 149–57. https://doi.org/10.1007/s10698-010-9086-5.

Lewis, Clarence Irwing. 1923. "A Pragmatic Conception of the *A Priori*." *The Journal of Philosophy* 20 (7): 169–77. https://doi.org/10.2307/2939833.

Lewowicz, Lucía. 2005. *Del Relativismo Lingüístico al Relativismo Ontológico en el Último Kuhn*. Montevideo: Universidad de la República.

Lombardi, Olimpia. 2011. "The Problem of Irreversibility, from Fourier to Chaos Theory: The Trajectory of a Controversy Space." In *Controversy Spaces: A Model of Scientific and Philosophical Change*, edited by Oscar Nudler, 77–102. Amsterdam: John Benjamins.

Lombardi, Olimpia. 2015. "The Ontological Autonomy of the Chemical World: Facing the Criticisms." In *Philosophy of Chemistry: Growth of a New Discipline*, Boston Studies in the Philosophy and History of Science, edited by Eric Scerri and

Lee McIntyre, 23–38. Dordrecht: Springer. https://doi.org/10.1007/978-94-017 -9364-3_3.

Lombardi, Olimpia, and Dennis Dieks. 2016. "Particles in a Quantum Ontology of Properties." In *Metaphysics in Contemporary Physics*, Poznan Studies in the Philosophy of the Sciences and the Humanities, edited by Tomasz Bigaj and Christian Wüthrich, 123–43. Leiden: Brill-Rodopi. https://doi.org/10.1163/9789004310827 _007.

Lombardi, Olimpia, and Martín Labarca. 2005. "The Ontological Autonomy of the Chemical World." *Foundations of Chemistry* 7 (2): 125–48. https://doi.org/10.1007 /s10698-004-0980-6.

Lombardi, Olimpia, and Martín Labarca. 2006. "The Ontological Autonomy of the Chemical World: A Response to Needham." *Foundations of Chemistry* 8: 81–92. https://doi.org/10.1007/s10698-005-9004-4.

Lombardi, Olimpia, and Martín Labarca. 2011. "On the Autonomous Existence of Chemical Entities." *Current Physical Chemistry* 1 (1): 69–75. https://doi.org/ 10.2174/1877946811101010069.

Lombardi, Olimpia, and Ana Rosa Pérez Ransanz. 2012. *Los Múltiples Mundos de la Ciencia. Un Realismo Pluralista y su Aplicación a la Filosofía de la Física.* México: UNAM-Siglo XXI.

Martínez González, Juan Camilo, Sebastian Fortin, and Olimpia Lombardi. 2019. "Why Molecular Structure Cannot Be Strictly Reduced to Quantum Mechanics." *Foundations of Chemistry* 21: 31–45. https://doi.org/10.1007/s10698-018-9310-2.

Marx, Karl. (1845) 1969. *Theses on Feuerbach*: *Marx/Engels Selected Works*, 13–15. Moscow: Progress Publishers.

Moulines, C. Ulises. 2004. "The Unity of Science and the Unity of Being: A Sketch of a Formal Approach." In *Logic, Epistemology, and the Unity of Science*, edited by Shahid Rahman, John Symons, Dov M. Gabbay, and Jean Paul van Bendegem, 151–61. Dordrecht: Springer. https://doi.org/10.1007/978-1-4020-2808-3_9.

Niiniluoto, Ilkka. 1999. *Critical Scientific Realism*. Oxford: Oxford University Press.

Pihlström, Sami. 2008. "How (Not) to Write the History of Pragmatist Philosophy of Science?" *Perspectives on Science* 16 (1): 26–69. https://doi.org/10.1162/posc .2008.16.1.26.

Pihlström, Sami. 2012. "Toward Pragmatically Naturalized Transcendental Philosophy of Scientific Inquiry and Pragmatic Scientific Realism." *Studia Philosophica Estonica* 5 (2): 79–94. https://doi.org/10.12697/spe.2012.5.2.06.

Primas, Hans. 1983. *Chemistry, Quantum Mechanics and Reductionism: Perspectives in Theoretical Chemistry*. Berlin: Springer. https://doi.org/10.1007/978-3-642 -69365-6.

Putnam, Hilary. 1981. *Reason, Truth and History*. Cambridge: Cambridge University Press. https://doi.org/10.1017/CBO9780511625398.

Putnam, Hilary. 1987. *The Many Faces of Realism*. La Salle: Open Court.

Quine, Willard Van Orman. 1981. *Theories and Things*. Cambridge, MA: Harvard University Press.

Sklar, Lawrence. 1974. *Space, Time, and Spacetime*. Berkeley, CA: University of California Press.

SPSP. 2019. *Society for Philosophy of Science in Practice: Mission Statement.* Accessed August 2019. https://www.philosophy-science-practice.org/about/mission-statement.

Torretti, Roberto. 2000. "'Scientific Realism' and Scientific Practice." In *The Reality of the Unobservable: Observability, Unobservability and Their Impact on the Issue of Scientific Realism*, Boston Studies in the Philosophy and History of Science, edited by Evandro Agassi and Massimo Pauri, 113–22. Dordrecht: Springer. https://doi.org/10.1007/978-94-015-9391-5_6.

Torretti, Roberto. 2005. *Manuel Kant.* Santiago de Chile: Ediciones Universidad Diego Portales.

Vihalemm, Rein. 2003. "Natural Kinds, Explanation, and Essentialism in Chemistry." *Annals of the New York Academy of Sciences: Chemical Explanation: Characteristics, Development, Autonomy* 988 (1): 59–70. https://doi.org/10.1111/j.1749-6632.2003.tb06085.x.

Vihalemm, Rein. 2005. "Chemistry and a Theoretical Model of Science: On the Occasion of a Recent Debate with the Christies." *Foundations of Chemistry* 7: 171–82. https://doi.org/10.1007/s10698-005-0959-y.

Vihalemm, Rein. 2011a. "The Autonomy of Chemistry: Old and New Problems." *Foundations of Chemistry* 13: 97–107. https://doi.org/10.1007/s10698-010-9094-5.

Vihalemm, Rein. 2011b. "Towards a Practical Realist Philosophy of Science." *Baltic Journal of European Studies* 1 (1(9)): 46–60.

Vihalemm, Rein. 2012. "Practical Realism: Against Standard Scientific Realism and Anti-Realism." *Studia Philosophica Estonica* 5 (2): 7–22. https://doi.org/10.12697/spe.2012.5.2.02.

Vihalemm, Rein. 2013a. "What Is a Scientific Concept: Some Considerations Concerning Chemistry in Practical Realist Philosophy of Science." *The Philosophy of Chemistry: Practices, Methodologies and Concepts*, edited by Jeanne-Pierre Llored, 364–84. Cambridge: Cambridge Scholars Publishing.

Vihalemm, Rein. 2013b. "Interpreting Kant's Conception of *Proper Science* in Practical Realism." *Acta Baltica Historiae et Philosophiae Scientiarum* 1 (2): 5–14. https://doi.org/ 10.11590/abhps.2013.2.01.

Westphal, Kenneth R. 2004. *Kant's Transcendental Proof of Realism.* Cambridge: Cambridge University Press. https://doi.org/10.1017/CBO9780511584497.

Wittgenstein, Ludwig. 1921. "Logisch-Philosophische Abhandlung." *Annalen der Naturphilosophie*, XIV. Translated by C. K. Ogden in Wittgenstein, Ludwig. 1922. *Tractatus Logico-Philosophicus.* London: Routledge & Kegan Paul.

Chapter 5

Practice-Oriented Realism in the Tradition of Rein Vihalemm

Hasok Chang

In this chapter, I propose to build on Rein Vihalemm's practical realism in three different ways. First, I rework some fundamental philosophical notions such as truth and reality by making them rooted in practices or activities. I believe this is fully in the spirit of Vihalemm's practical realism, or more broadly, what I will call "practice-oriented realism." Second, I show how various practice-oriented notions can serve as useful *tools of analysis*, rather than just being *subjects of debate*. This way we can try to avoid the futility of armchair pragmatism and put practical realism really into practice. In particular, I show how notions of epistemic activity and system of practice can serve as key framing devices in the historiography of science, through the case of my current work on the history of "battery science." Finally, I make a further development of the key notion of "operational coherence" in my thinking and discuss the process of "aim-oriented adjustment," through which our activities can become more coherent and serve as grounds for better truth and reality.

PRACTICE-ORIENTED REALISM

With his practical realism, Rein Vihalemm advanced notions of truth and reality that are suited to the understanding of scientific practices. I have tried to build on his insights in my own way, as part of my attempt to introduce pragmatist thinking more seriously into the philosophy of science. My thoughts in this direction have now been published in the form of a book titled *Realism for Realistic People: A New Pragmatist Philosophy of Science* (Chang 2022b). An early systematic exposition of the ideas developed in that book was given at the University of Tartu in March 2017, as the "Workshop on Pragmatist Realism: Philosophy, History and Science," consisting of five lectures, five classes for core participants, and a special laboratory practice session.[1] The Tartu workshop was a truly formative occasion, and at the conference in honor of Vihalemm in August 2019, I had the pleasure of returning to Tartu to present some further developments in my thinking. Much of the detailed substance of the talk I gave on that occasion has also been worked into the text of *Realism for Realistic People*. In the present chapter, I will try to avoid a direct repetition of the material that has been published, only giving a brief summary of it. On the other hand, I will add some further pertinent reflections that were not covered in my 2019 presentation in Tartu.

Let us start with a quick review of some highlights of Vihalemm's practical realism. The main tenets of practical realism are nicely summarized by Endla Lõhkivi and Rein Vihalemm (2012, 3). The first tenet declares: "Science does not represent the world 'as it really is' from a god's-eye point of view."[2] This is further elaborated as follows: "Naïve realism and metaphysical realism have assumed the god's-eye point of view, or the possibility of one-to-one representation of reality, as an ideal to be pursued in scientific theories. . . . Representationalism, however, is a tricky view since we lack independent criteria for judging the accuracy of a representation." In the third tenet, there is an emphasis that "scientific research is a practical activity and its main form is the scientific experiment that takes place in the real world, being a purposeful and critical theory-guided . . . material interference with nature." The fourth tenet notes that "science as practice is also a social-historical activity," and the fifth tenet explains that practical realism is "certainly realism, as it claims that . . . science as practice is a way in which we are engaged with the [real] world."

As I see it, Vihalemm's fundamental idea behind practical realism is that all knowledge, or even all discourse, is rooted in practice:

> According to practical realism, to speak about the world outside practice means to speak about something indefinable or illusory; only through practice can the objective world exist (as something specific and definite) for human beings. Therefore, knowledge must be regarded as the process of understanding how the world becomes defined in and through practice. (Vihalemm 2013, 9; see also Vihalemm 2012, 10)

And what exactly is practice? This is a deep question to which I cannot do justice here, but Vihalemm's brief characterization of it gives us a good start: "human activity as a social-historical, critically purposeful-normative, constructive, material interference with nature and society producing and reproducing the human world—culture—in nature" (Vihalemm 2012, 10).

Vihalemm's practical realism was a pioneering instance of what I will call practice-oriented realism, a category that encompasses a variety of positions including pragmatic realism, internal realism, and perspectival realism. In my own version of practice-oriented realism, the overall framework is the episte-mology of "active knowledge," namely knowledge taken as ability—know-ing how to do things, in the context of purposive action.[3] Active knowledge is something inherent in practices. Propositional knowledge, which is what ana-lytic philosophers have traditionally focused on in their treatment of knowl-edge, is only a particular and narrowly defined kind of knowledge. Without active knowledge there could not be any propositional knowledge (imagine we did not have the ability to learn and use language), and the importance of propositional knowledge would also be severely diminished if it did not have the function of facilitating active knowledge.

In Vihalemm's view, it is not only discourse that is grounded in practice, but objects themselves, too. This is a very important point in relation to how his position can qualify as "realism." Vihalemm (2012, 14) quotes Joseph Rouse (1987, 163) approvingly in this connection: "Belonging to the realm of possible determinations open within our practices is constitutive of a thing's being a thing at all." Vihalemm's metaphysics is a radical one, though he emphasizes that it is still a realist position:

> The practice-based approach implies that practical activity has a more fun-damental status than the status of individual objects-things. Concrete deter-mination of the existence of individual objects in this case is determined by specifically defined activities in the context of which these objects-things appear as specific invariants. (Vihalemm 2012, 13)

As his foil, Vihalemm presents "standard scientific realism," a position that has four key tenets: there is a "mind-independent world (reality)"; it is possible to obtain knowledge about it; truth is "correspondence between scientific statements (theories) and reality"; and truth is "an essential aim of scientific inquiry" (Vihalemm 2012, 10). He makes it clear that he objects to both standard scientific realism and antirealism. The main problem with standard scientific realism is that it is "isolated from practice," and presumes the God's-eye point of view. Antirealism, specifically in its empiricist–instru-mentalist or its (social) constructivist varieties, does avoid the latter problem, but not the former. This is because antirealism, too, arises from the context of "traditional philosophy of science, which is language- and logic-centered and does not proceed from the practice of real science" (Vihalemm 2012, 10–11).

Vihalemm's own view is anti-representational, if representation is taken in terms of correspondence to "self-identifying objects" (about which I will say more below). But if we reject the idea of truth as correspondence, what can truth be? I am not sure what the practical realist account of truth is, with Vihalemm's rejection of both pragmatism and the correspondence theory of truth. This is one place where I hope my own work can make a constructive contribution to Vihalemm's vision of philosophy. I am not sure if Vihalemm himself would have endorsed the moves I make since they involve embracing pragmatism, but I hope that pragmatism as I understand it[4] is something that practical realists can accept without going against Vihalemm's spirit.

So, how do I want to define truth? First of all, I think we can save ourselves a lot of trouble and hot air if we start by recognizing that the term "truth" (or "true") has multiple meanings.[5] One crucial distinction to recognize is between what I will call "primary" and "secondary" truth. Secondary truth consists in the agreement that a proposition has with other propositions that are already established to be true. The justification of a secondary truth can happen in various ways. "My cat is hungry" would be true by deduction if "All cats are hungry" were true, and it would be true by inference to the best explanation if the observable facts (that are already confirmed as true) can be accounted for better by the assumption that the cat is hungry than by any competing assumption. Now, this notion of secondary truth can incorporate something that deserves the designation of "correspondence" or "representation." If all the temperature measurements we have on record are true, then global warming is true. In such cases, our theory can be said to correspond to facts.

But this cannot be the whole story about truth. We must inquire why the ground of secondary truth (or, what makes it true) is already true. This is the inevitable challenge for any foundationalist enterprise. If our mode of justification is "A is true because B is true," there are only two ways in which the chain of justification can ultimately go: an infinite regress ("turtles all the way down"), or a vacuous kind of circularity. The latter is basically the idea that is often dignified with the name of the "coherence theory of truth," which has the inescapable problem that any self-consistent body of nonsense that someone might come up with should be considered true. Therefore, foundationalists have sought to identify some self-justifying or self-evidently true foundation of knowledge—but what could that be? Empiricists have tried the idea of indubitable sense-data, rationalists have tried the idea of indubitable metaphysical principles, and we know that both attempts have failed. Many standard scientific realists do something even more unrealistic and implausible, which is to hope that the chain of justification will end when we have somehow found the real picture of reality and confirmed that our theories correspond to that real picture.[6]

What we have to do is take a fresh look at the kind of statements that we do take to be so secure that they can serve as the foundation for other true statements, and ask why we regard them to be true. Take some examples:

- "This is my mother."
- "Here is a hand." (G. E. Moore)
- "The ground is firm."
- "The people that I see when I walk into a room are really there."
- "When I wake up in the morning, the earth will still be here."
- "A physical quantity is single-valued."
- "P v ¬P"
- "1 + 1 = 2"

In thinking about the truth of such statements, I try to partake in the spirit of what Wittgenstein (1969, § 471) said in a fragment included in the posthumous collection *On Certainty*: "It is so difficult to find the beginning. Or, better: it is difficult to begin at the beginning. And not try to go further back." Why do we take these statements as unquestioned truths? It is because the way we live is reliant on them, until and unless unexpected circumstances force us to change the way we live. I remember experiencing a major earthquake in October 1989 near San Francisco. For a short while afterward, all aspects of life were different (even just walking around), not being able to assume that the ground was firmly fixed. That strange existence stopped after a while, but that would have been different if major earthquakes had kept happening every few days. But as long as the reliance on the statement in question supports an effective way of life, then we go on regarding it as true. These thoughts point to a pragmatist conception of truth when it comes to "primary truth."

Let me try to make the ideas more precise now. This is the proposal I have made for a definition of primary truth in empirical domains: "A proposition is true to the extent that there are operationally coherent activities that can be performed by relying on it" (Chang 2022b, 167). I have dubbed this "truth-by-operational-coherence," and in the last section of the paper I will say more about what operational coherence means. The truth-by-operational-coherence of a proposition consists of the significant roles that it plays in our various activities. As William James put it, "Truth happens to an idea" as the idea goes on to facilitate further coherent activities (quoted in Kitcher 2012, xxiii). And we must admit that even the establishment of secondary truths involves operational coherence because checking a proposition against other propositions is itself an activity, based not only on logical principles but also on other fundamental assumptions such as the "principle of single value,"[7] which demands that a physical quantity does not take multiple values in a given situation. Truth conceived in this way is a quality with degrees and various dimensions, not a yes/no binary. That may sound strange if you are steeped in what you learned in elementary logic class, but it should not be offensive to most scientific realists, who are happy (or forced) to admit that even our best scientific theories are only "approximately" or "partially" true.

Another key concept that philosophers can hardly do without is reality, especially if we are claiming to be "realists." Vihalemm has made some suggestive remarks on this subject, which I will try to build on. I think he was right to say that "the world . . . does not consist of self-identifying objects" (Vihalemm 2013, 9), going against the traditional metaphysical realist view that "external reality" consists of objects that exist as they are regardless of how we conceptualize them, yet whose characteristics are exactly expressible in terms of the concepts that we possess (cf. Roberto Torretti, quoted in Chang 2022b, 7). Vihalemm's view is that reality only meaningfully exists within our practices:

> To speak about the world outside practice means to speak about something indefinable or illusory. It is only through practice that the objective world can really exist for humans. Therefore, knowledge must be regarded as the process of understanding how the world becomes defined in practice. (Vihalemm 2012, 10)

This sounds straightforward (if controversial), but there is a puzzle concerning Vihalemm's position on reality, as he also repudiated Putnam's internal realism and stopped short of endorsing Pihlström's pragmatist Kantianism, two positions that seem to me very much in line with practical realism.

I believe that there is a good way of reconciling Vihalemm's position on reality with pragmatist views. I suspect that Vihalemm wanted to preserve the idea of the mind-independence of reality, and he is not alone in this. What makes most people uneasy about pragmatist-leaning notions of reality is the worry that if we follow such ideas we would lose mind-independent reality: then the tree would not fall in the forest if we were not there to hear it, and so on, and we would be creating reality by our thought. But I believe that this worry is based on a confusion. It is helpful to disambiguate the notion of mind-independence into what I call "mind-framing" and "mind-control." Everything we can ever talk or think about is *mind-framed*—in other words, conceived in terms of some concepts that come from our mind (that is to say, not "self-identifying," in Vihalemm's idiom). But that does not mean that such entities will do as we wish; in other words, they are *not mind-controlled*. We ask the questions, and nature gives the answers. We may have our wishes, but nature will often resist them. I would say that real entities are mind-framed and mind-uncontrolled things.

In my recently published book, I gave the following definition of what it means to be "real": "an entity is real to the extent that there are operationally coherent activities that can be performed by relying significantly on its existence and its properties" (Chang 2022b, 121). So, operational coherence grounds reality as well as truth. This is intended as a definition of reality[8] that can serve to explicate what we mean by "reality" in ordinary discourse. And it is important to note that my pragmatist definition of reality (and truth as well) is not subjectivist: whether there are operationally coherent activities that can be performed with a certain object (or on the basis of a certain proposition) is not dependent on whether I can personally perform them or whether I personally know about them. For example, as I have argued elsewhere (Chang 2022b, 151–53), phlogiston has a measure of reality because there are a set of coherent activities that one can perform with it, which a community of chemists in the eighteenth century knew how to perform. The reality of phlogiston does not cease just because that phlogistonist community has died out and their knowledge has been forgotten. Historians of science today can recover (and have recovered) the knowledge of that reality.

When we say that the placebo effect is real, what we really mean is that there are coherent things that we can do with the placebo effect (e.g., relieve someone's symptoms by letting it be, or control for it through randomization in clinical trials). When we say that the Loch Ness Monster is not real, we mean we cannot do much with it that is coherent—has anyone been able to capture it, ride it, get a sample of its DNA, or even get a clear photograph of it? Sure, we can gossip and mythologize about it, but there is far less that we can do with the Loch Ness Monster than with other things that we can get a grip on as existing material objects, including the dinosaurs whose fossilized skeletons we can study and display. I submit that this is the kind of conception of reality that practical realists should subscribe to, because it is fully embedded in practices. Operational coherence is *not* an *indication* of existence that is predetermined in some other way; rather, it is *constitutive* of something's being real. There are some clear parallels between this pragmatist notion of reality and the pragmatist notion of truth that I discussed above. Reality also becomes a matter of quality with degrees and various dimensions, not a yes/no binary.

What about realism, then? What can practice-oriented realists claim about truth and reality? There are two main things to say. First, practice-based notions of truth and reality allow us to make a credible claim that science does provide us with various truths about various realities. However, that claim, by itself, would also give us an impoverished view of what science is all about. One of the most distinctive and essential characteristics of science as a practice is that it is driven by the striving to have more and better knowledge. Realism should not be a passive and merely descriptive doctrine. A proper "ism" should be an ideology, a commitment to a way of life or at least a set of values. To emphasize this aspect of science, I advocate "activist realism," which is "a commitment to do whatever we can in order to extend and enhance our knowledge concerning realities, as much as possible in the context of other aims and values" (Chang 2022b, 209). This becomes a realistic ideal when reality is conceived in the practice-based way that I have proposed in Vihalemm's spirit. And there is no reason why those whose positions are commonly labeled "antirealist" should object to activist realism, as long as they are in favor of the expansion and enhancement of empirical knowledge; the pragmatist renditions of the notions of truth and reality serve to remove the legitimate worries about standard scientific or metaphysical realism that thoughtful antirealists have raised.

PRACTICE-ORIENTED REALISM IN HISTORIOGRAPHICAL ACTION

Why should anyone be a practice-oriented realist? One might even ask: What is the practical use of practical realism? This is not an unfair question. As a pragmatist, I believe that philosophy itself should be judged through its uses; my pragmatism is reflexively pragmatist about the function of philosophy. Practice-oriented philosophies of science can certainly help us reach a proper and well-rounded understanding of real science, science as it is actually practiced, or "science in practice." This is a task of crucial importance in our modern scientific-and-technological age. What goes by the name of "philosophy of science" needs to step up to the challenge of providing useful conceptual tools for all people who want to understand the nature of science and engage with science in various ways. My work is intended to make a philosophy of science that is fit for understanding and even facilitating actual scientific practices, actual scientific knowledge, and actual scientific progress. If that works out well, then that will be a vindication of my ideas. This is also in the spirit of "conceptual engineering," in the apt expression that is becoming popular these days.

The understanding of science in practice includes the understanding of science as it was practiced in the past—in other words, the historiography of science. An important use of the philosophy of science is the framing of historical accounts of the development of science. This is an important concern in my own work, which is strongly focused on "integrated history and philosophy of science." To those who may not value this academic use of the philosophy of science, let me just say: How often is it that abstract philosophical ideas are useful, even for another academic discipline? And giving such academic aid can have great practical importance, too, if the academic discipline aided by our philosophy has practical consequences. I believe this is indeed the case for the field of history of science (and science and technology studies more broadly).

There are many useful suggestions for the philosophical framing of the history of science that can come from practice-oriented philosophy. The most basic one would be to understand knowledge as a matter of ability, not merely as the possession of information. There is also the suggestion to take truth and reality as concepts based on the operational coherence of activities, as explained above. Such philosophical reorientations have a significant impact on how we view the history of science. Of more immediate effect, however, is the attention to scientific practice itself. In my own work as a historian of science, I have made much use of the notions of "epistemic activities" and "systems of practice" as units of analysis. It is worth noting that I did not generate these historiographical concepts in a "top-down" way from a preconceived pragmatist philosophy. Rather, they arose by necessity in my struggle to find better ways of conceptualizing what was going on in the past science that I was analyzing, particularly in the set of studies in eighteenth- and nineteenth-century chemistry and physics that were published in my book titled *Is Water H$_2$O?* (Chang 2012).

In that publication, I defined the concepts as follows (Chang 2012, 15–16; see also Chang 2014; Chang 2022b, section 1.3). An epistemic activity is "a more-or-less coherent set of mental or physical operations that are intended to contribute to the production or improvement of knowledge in a particular way, in accordance with some discernible rules." And a system of practice is "formed by a coherent set of epistemic activities performed with a view to achieve certain aims." Already implicit in these characterizations was the notion of operational coherence (which I simply called "coherence" there), which I was able to articulate only much later. It may seem uncontroversial to think of scientific practice in these terms, but it is very far from the traditional inclinations of philosophers of science, who would normally think in terms of propositions (including experimental results), and other propositional entities (such as theories and theoretical models). If we are limited to a propositional framing of science, we will miss a great deal about what goes on in scientific practice, including the acquisition of observational and computational *capabilities*, and the kind of hands-on interventions that Ian Hacking (1983) was fond of highlighting. In recent times, many philosophers of science have found the traditional propositional framing of science frustrating and have sought to go beyond it. This newer trend is especially evident in the work promoted by the Society for Philosophy of Science in Practice (of which I was a cofounder).

I will now try to illustrate, through one extended example, the benefits of using the notion of systems of practice (and, by implication, epistemic activities) as a framing device in the historiography of science. The example is my current work on the history of batteries, and of what I have called "battery science" in the nineteenth century. If you know anything at all about this history, you probably know about the sensation caused by Alessandro Volta's invention of the "pile" (announced in 1800) and its productive employments—starting with enlightened entertainment, but soon including scientific uses such as electrolysis, and important practical uses such as powering the global telegraph network. And without batteries, there would have been no electric circuits, without which there would have been no discovery of the electromagnetic effect, without which there would have been no electric motors or generators, without which we would not have the electric power grid (and could not do any of the things that we take for granted when we plug our devices, including rechargeable batteries, into the outlet in the wall). And imagine trying to do any scientific research nowadays without an electricity supply. So, it is not an exaggeration to say that there would have been no scientific and technological civilization as we know it, without the invention of the battery and all that followed from it. It is a crucially important subject that has not received very much attention from historians, though there are some notable exceptions including works by Richard Schallenberg (1982) and James Morton Turner (2022).

Now, if you know a little more about the history of batteries, you probably also know about the long-standing dispute that took place about how it is that batteries work. It was very easy to rig up a battery as described by Volta: put down on the table a piece of silver, and on top of it a piece of zinc. On top of the zinc piece, put a piece of paper (or anything else wettable) soaked in water, or better, in salt water. Put on top of the wet layer another piece of silver, and a piece of zinc, and another wet layer. Repeat. One can also use various other kinds of metals. People could make this device even from what was briefly reported in a daily newspaper, and other sketchy descriptions. But it was a whole other matter to explain how this contraption generated a flow (or accumulation) of electricity. Volta himself thought that it was the contact between two different metals that caused the movement of the electrical fluid, which the wet layer merely conducted. Volta had plenty of followers, but soon many others advanced an opposing view. According to that view, what excited the flow of electricity was the chemical reaction between one of the metals (zinc in this case) and the wet stuff (the electrolyte, as it came to be called later), with the other metal merely serving as a conductor. A long controversy ensued.

The dispute between the "contact theory" of Volta and the "chemical theory" of his opponents has provided the most common way in which the early history of batteries has been framed by historians of science. Here is one example of history written in that framing, by Helge Kragh, that I actually admire a great deal: "What makes the voltaic controversy both interesting and unusual is its long duration and complex structure. . . . [N]one of the great theoretical breakthroughs of the [19th] century . . . had a decisive influence on the controversy." Kragh points out that the controversy even lasted into the twentieth century: "although, in a social sense, the controversy had largely disappeared by 1910 there was no consensus on the question of whether or not contact potentials exist as an intrinsic property of metals" (Kragh 2000, 153).

All of that is correct, but there are significant limitations to this way of framing the history. The focus on theoretical dispute is a natural consequence of the propositional framing, which directs attention almost exclusively to theories and their assessment in relation to observational reports. For one thing, the propositional framing cannot take in developments in instrumentation very well. In an area such as battery science it is impossible to ignore the material instruments, so we end up treating theoretical debates and instrumental developments quite separately, the latter going into the realm of the history of technology. Focusing on the theoretical dispute also tends to result in too much attention on winning and losing, because a natural presumption in the propositional framing is that a proposition is either true or false. This gets in the way of paying attention to the distinct achievements made by each side. (There is also a problem in only seeing two sides to the development, when there were other traditions of work, too.)

To remedy these shortcomings, I reframe this history through the kind of pragmatist reorientation outlined above. Instead of competing *theories*, I have identified *systems of practice*, and as many as four of them, which developed in parallel and in interaction with each other. A system of practice incorporates not only theories and concepts but experiments, instruments, and technological developments; they include various kinds of activities with various aims, more or less coherent each in itself, and in relation to each other and with respect to broader aims. I trace the development of knowledge achieved in these systems, tracking the abilities acquired, the truths learned, and the realities identified and created within each system. And I also note how the different systems interacted productively with one another, especially because there were certain problems that could not be solved effectively in any one system. My forthcoming book titled *How Does a Battery Work?* will contain extended accounts of how knowledge grew in each system. For now, here is a very brief summary.

The *contact–electrostatic system* was initiated by Volta himself, who identified the *contact* between different metals (or other substances) as the seat of electrical action, and conceived the effects in terms of eighteenth-century *electrostatics* that reasoned in terms of the distribution of the electrical fluid over different bodies. In today's orthodox electrochemistry Volta's view on the "contact potential" is firmly rejected, or more often, simply ignored. Yet the system of practice that arose from Volta's work developed coherently and produced some impressive results. These included two iconic instruments. The first was the "dry pile," a voltaic pile containing no wet layers and therefore, presumably, no chemical reactions. The other was the thermocouple, a simple instrument consisting of two pieces of wire made of different metals, connected to each other at both ends to make a loop; when the two junctions are put at different temperatures, a voltage difference arises, and current flows. The thermocouple was the battery used in Georg Ohm's experiment that established Ohm's law and led to the firm establishment of the very concepts of voltage, current, and resistance. After Ohm's work, the contact–electrostatic system seemed to stagnate, but then William Thomson (Lord Kelvin) brought new life to it in the 1860s by his invention of the "quadrant electrometer" which was able to detect the voltaic contact potential produced by a single bimetallic pair. And in the early twentieth century the new experimental physics of electrons gave rise to various clear manifestations of the voltaic contact potential observed in the photoelectric effect, thermionic emissions, and the physics of semiconductors. Hear Irving Langmuir's reflections from 1916:

> Thus the long-standing conflict between the [contact and chemical theories of Voltaic action] was apparently finally brought to a close by the complete victory of the chemical theory. Within the last four or five years, however, some remarkable progress has been made in certain branches of physics, which has resulted in bringing to life again the contact potential theory. (Langmuir 1916)[9]

The contact–electrostatic system was not simply wrong, and Volta's contact potential has exhibited lasting reality.

The *chemical imbalance system* was the direct ancestor of today's orthodox electrochemistry. The basic idea there is that the flow of electricity is generated by lopsided chemical action. A voltaic cell consists of two "half-cells," in each of which a characteristic chemical reaction takes place; one side has a stronger electrochemical action than the other, and the overall direction of electrical current is determined by the imbalance. Later, this idea was formulated in terms of reduction-oxidation (redox) potentials. The iconic instrument of this system was the Daniell cell, which is what nearly all of today's chemistry textbooks present when they try to explain how batteries work. Originally invented by John Frederic Daniell in the 1830s as the "constant battery," the Daniell cell was originally an instrument of merely practical importance. Daniell introduced a series of practical modifications to the then-standard configuration of the voltaic cell, which used pots of electrolyte rather than Volta's solid layers soaked in the electrolyte. The effective configuration reached by Daniell's practical adjustments was a two-metal two-electrolyte arrangement, in which each metal piece was dipped in a solution containing its own ions. Why this was a good configuration for making a battery with a steady output was not clear. Only decades later, the theoretical recognition came that a Daniell-type half-cell perfectly embodied the redox potential of a metal, and with that development, the Daniell cell became the iconic textbook battery, embedded in the theoretical framework of chemical thermodynamics and physical chemistry. Ironically, this was just when it was nearing the end of its life as an important practical battery, eclipsed by newer and more useful instruments.

The *conservationist system* was what gave rise to what many people would now consider the most important type of battery: namely, the rechargeable batteries in so many of our devices these days, including mobile phones, laptops, electric cars, and facilities for generating and storing renewable energy. Rechargeable batteries were invented very early on, but they did not have much practical significance for a whole century. They came to play an important conceptual role, however, as a convincing illustration of the interconversion of different types of energy. In fact, batteries were a very important part of Hermann Helmholtz's grand synthesis of science under the umbrella of the conservation of "force" (energy) in 1847. Note that considerations of energy were absent for many decades in the first two systems of battery science discussed above. William Robert Grove, whose thinking was strongly influenced by his rather accidental invention of a rechargeable gas battery, was identified by Thomas Kuhn (1977) as one of the independent discoverers of the principle of energy conservation. Grove's gas battery is basically a setup in which the electrolysis of water into hydrogen and oxygen gases can be run backwards with the help of a platinum catalyst; it is now recognized as the first-ever hydrogen fuel cell, though no one at the time saw it as having any practical use.

The *corpuscular–mechanical system* of battery science was driven by the desire for microphysical explanations, which were not given very often in any of the three other systems until the end of the nineteenth century. Corpuscular–mechanical thinking in this field was initially spurred on by the desire to understand electrolysis (the decomposition of chemical substances by means of a battery), which was perhaps the first significant new phenomenon produced by the voltaic pile. William Nicholson and Anthony Carlisle made an electrolysis of water into hydrogen and oxygen almost immediately upon hearing about Volta's invention and making their own voltaic pile. The effect was marvelous, but it raised a serious puzzle. It is easy enough to imagine that the electrical action of the voltaic pile somehow broke up a molecule into its constituents, but how could one explain the fact that the broken-up constituents appeared separately at the two electrodes, at a macroscopic distance from each other?[10] Some microscopic story had to be given, and many hypotheses and models arose. Thinking about electrolysis also led to the electrostatic theory of chemical bonds by Humphry Davy and Jöns Jacob Berzelius. More credible theorizing about ions and experimental tracking of them became possible in the late nineteenth century, and this was an essential strand of work going into the establishment of the whole field of physical chemistry.

Even though I have only given an extremely sketchy view of my work on the history of battery science here, I hope that it is suggestive enough of the advantages of the practice-oriented framing of the historical developments. Consider how impoverished the historiography of battery science would be (and has been), either with an exclusive focus on theories and theoretical models and their justification, or with compartmentalized treatments of theory, experiment, and instrumentation,[11] and of science and technology.

As with the case of the origin of my notions of epistemic activity and system of practice, the framing of the history of battery science is much more than an *illustration* of my pragmatist philosophical ideas. Rather, the philosophical ideas used in the framing of the history actually *arose* in the course of my struggle to find a good framework for telling the story of battery science. More generally, my philosophical ideas should be fit to serve the purposes of historiography because they are usually *forged* in the course of my historical work.

OPERATIONAL COHERENCE AND ITS IMPROVEMENT

So far, I have made much use of the notion of "operational coherence" without giving a precise characterization of it. In this section, I will attempt to say more about what operational coherence is, and how it is that we go about improving it as we seek to advance knowledge. I believe that something like operational coherence is an essential notion for all practice-oriented realists, because it is about the quality of our activities—and practices are activities, whatever else they might be. It does not make sense to say that an activity is successful because it somehow corresponds correctly to the world. Rather, an activity works out because, roughly speaking, what goes into it all fits together well. It is important to note that coherence as I intend it here is about the harmoniousness of actions, not primarily about the logical relationship between propositions. It is in order to mark that point clearly that I use the phrase *"operational* coherence."[12]

Operational coherence is a concept that I have been developing over the course of more than a decade now, and it has had many renditions.[13] A succinct formulation I have arrived at by now, which I think will be quite stable, is that operational coherence is *aim-oriented coordination* (Chang 2022b, 40). Another way to see it is that an activity is operationally coherent if it is designed well for the achievement of its aim.[14] Operational coherence may consist in something as mundane as the correct coordination of bodily movements and material conditions needed in riding a bicycle, lighting a match, walking up and down the stairs (or walking at all), or drinking a glass of water. Or it may be something as esoteric as the successful integration of a range of material technologies and a collection of abstract theories in the operation of the global positioning system. In all of these cases, we can meaningfully ask the question of good activity-design.

It is important to note that operational coherence is not a mind-independent quality; there is a clear hermeneutic dimension to operational coherence. One way to put this point is to say that operational coherence is all about *doing what makes sense*. The coherence–success relation is not one of cause and effect, but a hermeneutic and pragmatic act of sense-making. If we are discussing success we are in the realm of actions, in the context of goal-oriented behavior. So, what is operationally coherent is what makes sense for us to do, and the "sense" here is framed in terms of aims. It is not that an activity is successful *because* it is coherent, but that the coherence of an activity *consists* in doing what makes sense. Coherent activities tend to work because they are carefully designed so that they would work. And sense-making here is certainly not just "in the head"—it is about reaching a combined bodily and conceptual understanding of what we do. Such understanding is built through a process of incorporating activities-that-work into our conceptions (e.g., think back to the case of the Daniell cell discussed in the last section). This also implies that coherence does not pertain to a single act but to a sustained and organized *activity*. And a system of practice, which consists of a coherently coordinated set of activities, is a whole regime of understanding.

As a brief initial illustration of this sense-making, consider the classic example of bicycle-riding. The tacitness or inarticulateness of the skill involved in this activity, so memorably emphasized by Michael Polanyi, is very apt for my discussion here. Initially, the novice does not understand how to keep himself from falling. The helpful older sister gives him tips like "turn into the direction in which you are beginning to fall." This advice makes no sense in the abstract, and when the boy tries to put it into practice, it does not work. However, trial and error eventually shifts something in the brain and the muscles, and he is able to ride, wobbly as he may be. At this stage, the thing about turning into the direction of falling starts to make that conceptual–bodily sense. As his skill improves, he also begins to understand things like how a slight turn can be achieved by the shifting of body weight without turning the handle. Through such improvements, his bicycle-riding continues to improve its coherence.

To understand coherence, it also helps to think about cases of incoherence. Some incoherence can be purely mental. In colloquial usage, we often say "incoherent" when someone is talking gibberish, not understandable, "not even wrong." But it is more interesting and informative to consider activities whose incoherence consists in an ineffective conjunction between how we think, what we want, what we do, and the way things are. For example, I cannot practice archery coherently without good coordination of my own strength, the properties of the bow and the arrow and the surrounding air, the location of the target, and the basic laws of mechanics. An occasional failure may be explained away ("oops, my hand slipped as I was stretching the bow," or, "damn, there was this sudden gust of wind that I hadn't expected"), but if I keep missing the target completely and do not have good explanations of my failures, then we would be tempted to say that my whole activity is incoherent.

The consideration of incoherence brings me to a very important issue: What is the process through which we improve the operational coherence of our activities? This has a close relation to John Dewey's view on the nature of inquiry. For Dewey, what gives rise to inquiry is a disturbed situation. In my terms, a disturbed situation is a situation of operational incoherence, where our activities are not sufficiently successful and we have no plausible account of the failures as explainable exceptions. In that case, we must launch an investigation, in the way in which Dewey saw a disturbed situation giving rise to inquiry. When something is not working out, we must start thinking and doing things differently. Let us return to the example of archery for a moment. If I keep missing the target, maybe I have to pull the bowstring harder, or revise the way I assess the amount of force exerted, or get my eyes checked to see if I can see the target well enough, or even adopt different laws of physics. I must try one thing and another and then another, whatever it takes, until I am able to hit the target better.

What is illustrated through this example is a process that I call "aim-oriented adjustment."[15] If we just consider well-defined and simple activities for the moment, we can say that each activity has a clear and identifiable aim. The process of improving our activity is guided by its main aim. We adjust what we do, in order to meet that aim more effectively. Through this process our activity becomes more operationally coherent (or, better designed). There is no pre-given algorithm for a process of aim-oriented adjustment. We have to twist-and-turn, sometimes literally, until we are more comfortable (as in bed) or until something clicks (as in mechanical assembly). Such processes are prevalent in all kinds of learning processes, starting with the acquisition of skills in everyday life: hammering a nail, using chopsticks, lighting a match, or throwing a Frisbee or an American football. And then there are skills that are commonly recognized as challenging to learn, everything from juggling to making a beautiful resonant sound on a violin. It is also interesting to note that many of the skills pertain to second-person activities and require for their learning a joint effort in aim-oriented coordination.

Scientific practices require many of these everyday skills, and on top of them, many more advanced ones. And making scientific activities work requires a higher degree of aim-oriented coordination. This is similar to what Ludwik Fleck called "tuning," explained by Andrew Pickering (1995, 121) as follows: "The scientists tried varying the prototype recipe [for the Wasser-mann test for syphilis] in all sorts of ways and eventually arrived at a recipe that was medically useful." The process by which Daniell made his stable battery, briefly mentioned in the previous section, is another good example of aim-oriented adjustment in science. And in that case, the practical adjust-ment was only followed much later by the conceptual adjustment allowing a deeper understanding of why Daniell's instrument worked so well. Both steps were required for a compelling instance of aim-oriented adjustment, through which emerged a whole new articulation of the chemical imbalance system of battery science. Generally speaking, aim-oriented adjustment constitutes an important method of discovery and innovation in science, and attention to aim-oriented adjustment provides a way for the philosophy of science to engage sensibly with the context of discovery again.

Aim-oriented adjustment is an adaptive process, driven at each moment by the relief and satisfaction provided by improved coherence, without a fixed final destination. Anything and everything that is within our power to change may be changed in the process of increasing operational coherence. There is nothing in our activities that is fixed and validated forever and uncondition-ally. The same can be said about inquiry, in the Deweyan spirit. Inquiry does not follow predetermined and eternal methods, and it is a process of method-learning as well as content-learning. Now, it is certainly the case that a great deal of scientific work is done in a much more prescribed way, as Kuhn emphasized in his discussion of normal science, which follows specific and narrow pathways laid out by the dominant paradigm in a field. However, if we take the *whole* Kuhnian picture of the development of science, including the phases of extraordinary research and scientific revolutions, we can eas-ily recognize that there are no firm universal restrictions on what is open to adjustment. It makes sense to view inquiry in general as an unrestricted pro-cess of aim-oriented adjustment. And it also makes sense that under various circumstances, various temporary restrictions can fruitfully be placed on what may be adjusted in the process of inquiry.

An important aspect of the openness of aim-oriented adjustment is that even aims themselves are subject to change in the process. There are many reasons why we might want to change our aims, but two of them seem par-ticularly important. First, our aim can and should change if we learn that its achievement is not plausible. Then the best way to enhance operational coherence is to try for a different aim. Strictly speaking, changing the aim of an activity amounts to giving up on that activity and engaging in a different one. I retain the notion of aim-orientedness even in that situation, because the operational coherence of an activity is still defined in relation to the aim of the activity at each moment, even though the aim is not fixed in the long term. The change of aim based on achievability can happen at all scales. On a large scale, consider Bas van Fraassen's (1980) stance that science should not aim to attain truth about unobservable entities because that is not the kind of thing that science can actually achieve. On a small scale, we may keep failing in hammering a nail into a wall, only to discover that behind the wallpaper is a sheet of steel. The best thing to do in that circumstance is to give up on the aim of putting nails into that wall, because no plausible adjustment in our hammering activity is going to achieve that aim. If our larger aim is to hang a picture on the wall, then the best course of action is to give up on the nail, and perhaps look for a way to glue a hook on the wall.

The last consideration leads me to the second important reason for the change of aims, which is the fact that our activities interact with each other, and an activity can also be nested within another. The ontology of activities is complex, and this is particularly important when we consider systems of practice. A simple activity has one well-defined aim. In contrast, a system of practice will have system-level aims that cannot be boiled down to the aim of any activity that operates within the system. Aims of various activities can compete and clash with each other, and they must be moderated in order for the whole system of practice to maintain and improve its operational coherence. This is very easy to see in the context of politics or economics. The aim of extending liberty often conflicts with the aim of ensuring security. The aim of reducing inflation is laudable, but it must be given up if the means of achieving it result in the reduction of wages and unsustainable interest payments for borrowers, which will interfere with other aims we have in society. The adjustment of various competing and interacting aims is an important aspect of the art of communal life, and very similar considerations are essential in systems of scientific practice. To return to the main example from the previous section: all systems of battery science had a dual aim of producing new electrical phenomena and explaining them theoretically. Each system employed different instruments, concepts, and theories to these ends, and engaged in various specific experimental and theoretical activities, which had to be carefully coordinated with each other. Care needed to be taken to ensure that the functioning of one's favorite instrument was well understood by reference to one's favorite concepts. For that reason, practitioners of the contact–electrostatic system were not able to take full advantage of all the various chemical reactions that could make a basis for batteries, while practitioners of the chemical imbalance system were led to shun the use of thermocouples and other nonchemical batteries. Sometimes the aims clashed with each other. At a large scale, it often happened that if a system of practice produced a profusion of new phenomena, some of them were bound to be difficult to explain in its favored mode. In such cases it became necessary to restrict attention to selected phenomena, restraining the aim of producing as many phenomena as possible. Achieving operational coherence in a system of practice is a complex task demanding a highly sophisticated level of aim-oriented coordination.

CONCLUDING REMARKS

I hope I have been able to present some plausible directions of development that we can take from Rein Vihalemm's pioneering thoughts on practical realism. I believe that the substance of his philosophy, or at least its spirit, can usefully instruct all practice-oriented philosophy. I imagine that Professor Vihalemm would have approved of the pragmatist notions of knowledge, truth, and reality that I present here. In my opinion, there is no better way to advance philosophy than to pay serious and meticulous philosophical attention to the practices of life, including science.

NOTES

1. A similar exposition was given earlier in Bielefeld, and later in London and Edinburgh.
2. Vihalemm (2012, 10) attributes the phrase "god's-eye point of view" to Putnam (1990, 11).
3. See Chang (2022b, chapter 1).
4. My take on pragmatism, as a thorough and relentless kind of empiricism, is explained in Chang (2022b, section 1.6).
5. See Chang (2022b, section 4.2).
6. The postulation of "the world" to which our primary truths correspond is like the postulation of "God" as the source of our ethical rules and the divine right of kings.
7. See Chang (2008).
8. Note that the English term "reality" is grammatically ambiguous: it may mean "the quality of being real" or "a real thing." My definition of "real" can also accommodate this dual use.
9. See Chang (2022a) for further details on the curious life of the voltaic contact potential.
10. See Chang (2012, chapter 2) for a more detailed account of how nineteenth-century scientists tried to deal with this "distance problem" concerning electrolysis.
11. See Galison (1997) for an integrated history of high-energy physics, which emphasizes the interaction between these strands while clearly distinguishing them as quite autonomous.

12. In this phrasing, I am giving a conscious nod to Percy Bridgman's advocacy of the operational point of view; see Chang (2009) for further details on my interpretation of Bridgman's philosophy. And, as I will attempt to argue in later publications, I believe that my operational notion of "coherence" is implicit in John Dewey's theory of knowledge.

13. Chang (2017) was an important step in the process, although I have now significantly modified the formulations given there. That stage of development is also meaningful to look back on in relation to the present chapter, as it was essentially what I presented in the special workshop in Tartu in 2017.

14. In the 2017 formulation, I stated: "an activity is operationally coherent if and only if there is a harmonious relationship among the operations that constitute the activity; the concrete realization of a coherent activity is successful, *ceteris paribus*; the latter condition serves as an indirect criterion for the judgement of coherence" (Chang 2017, 111).

15. See Chang (2022b, section 1.5).

REFERENCES

Chang, Hasok. 2008. "Contingent Transcendental Arguments for Metaphysical Principles." In *Kant and Philosophy of Science Today*, Royal Institute of Philosophy Supplement 63, edited by Michela Massimi, 113–33. Cambridge: Cambridge University Press. https://doi.org/10.1017/S1358246108000076.

Chang, Hasok. 2009. "Operationalism." *Stanford Encyclopedia of Philosophy*. Accessed October 2023, https://plato.stanford.edu/archives/fall2009/entries/operationalism.

Chang, Hasok. 2012. *Is Water H_2O? Evidence, Realism and Pluralism*. Dordrecht: Springer. https://doi.org/10.1007/978-94-007-3932-1.

Chang, Hasok. 2014. "Epistemic Activities and Systems of Practice: Units of Analysis in Philosophy of Science after the Practice Turn." In *Science After the Practice Turn in the Philosophy, History and Social Studies of Science*, edited by Léna Soler et al., 67–79. London and Abingdon: Routledge.

Chang, Hasok. 2017. "Operational Coherence as the Source of Truth." *Proceedings of the Aristotelian Society*, 117 (2): 103–22. https://doi.org/10.1093/arisoc/aox004.

Chang, Hasok. 2020. "A Tale of Two Batteries." In *Rotas, Mapas & Intercâmbios da História da Ciência*, edited by Ana Maria Alfonso-Goldfarb et al. São Paulo: Educ.

Chang, Hasok. 2022a. "Dead or 'Undead'? The Curious and Untidy History of Volta's Concept of 'Contact Potential.'" *Science in Context* 34 (2): 227–47. https://doi.org/10.1017/S0269889722000199.

Chang, Hasok. 2022b. *Realism for Realistic People: A New Pragmatist Philosophy of Science*. Cambridge: Cambridge University Press. https://doi.org/10.1017/9781108635738.

Galison, Peter. 1997. *Image and Logic: A Material Culture of Microphysics*. Chicago, IL: University of Chicago Press.

Grove, William Robert. 1839. "On Voltaic Series and the Combination of Gases by Platinum." *The London, Edinburgh and Dublin Philosophical Magazine and*

Journal of Science (3rd Series) 14 (86–87): 127–30. https://doi.org/10.1080 /14786443908649684.

Hacking, Ian. 1983. *Representing and Intervening*. Cambridge: Cambridge University Press. https://doi.org/10.1017/CBO9780511814563.

Helmholtz, Hermann. [1847] 1971. "The Conservation of Force: A Physical Memoir." In *Selected Writings of Hermann von Helmholtz*, edited by Russell Kahl, 3–55. Middletown, CT: Wesleyan University Press.

Kitcher, Philip. 2012. *Preludes to Pragmatism: Toward a Reconstruction of Philosophy*. New York: Oxford University Press. https://doi.org/10.1093/acprof:oso /9780199899555.001.0001.

Kragh, Helge. 2000. "Confusion and Controversy: Nineteenth Century Theories of the Voltaic Pile." In *Nuova Voltiana: Studies on Volta and His Times*, Volume 1, edited by Fabio Bevilacqua and Lucio Fregonese, 133–57. Pavia: Università degli Studi di Pavia.

Kuhn, Thomas S. 1977. "Energy Conservation as an Example of Simultaneous Discovery." In *The Essential Tension: Selected Studies in Scientific Tradition and Change*, 66–104. Chicago, IL: University of Chicago Press.

Langmuir, Irving. [1916] 1961. "The Relation Between Contact Potentials and Electrochemical Action." In *The Collected Works of Irving Langmuir, Volume 3 (Thermionic Phenomena)*, edited by Chauncey Guy Suits and Harold E. Way, 173–217. New York: Pergamon. https://doi.org/10.1016/B978-0-08-009355-0 .50021-9.

Lõhkivi, Endla, and Vihalemm, Rein. 2012. "Philosophy of Science in Practice and Practical Realism." *Studia Philosophica Estonica* 5 (2): 1–6. https://doi.org/10 .12697/spe.2012.5.2.01.

Pickering, Andrew. 1995. *The Mangle of Practice*. Chicago, IL: University of Chicago Press.

Putnam, Hilary. 1990. "Realism with a Human Face." In *Realism with a Human Face*, 3–29. Cambridge, MA: Harvard University Press.

Rouse, Joseph. 1987. *Knowledge and Power: Toward a Political Philosophy of Science*. Ithaca, NY: Cornell University Press.

Schallenberg, Richard H. 1982. *Bottled Energy: Electrical Engineering and the Evolution of Chemical Energy Storage*. Philadelphia, PA: American Philosophical Society.

Turner, James Morton. 2022. *Charged: A History of Batteries and Lessons for a Clean Energy Future*. Seattle: University of Washington Press.

Van Fraassen, Bas. 1980. *The Scientific Image*. Oxford: Clarendon Press. https://doi .org/10.1093/0198244274.001.0001.

Vihalemm, Rein. 2012. "Practical Realism: Against Standard Scientific Realism and Anti-Realism." *Studia Philosophica Estonica* 5 (2): 7–22. https://doi.org/10.12697 /spe.2012.5.2.02.

Vihalemm, Rein. 2013. "Interpreting Kant's Conception of *Proper Science* in Practical Realism." *Acta Baltica Historiae et Philosophiae Scientiarum* 1 (2): 5–14. https://doi.org/10.11590/abhps.2013.2.01.

Wittgenstein, Ludwig. 1969. *On Certainty (Über Gewissheit)*. New York: Harper.

Chapter 6

Mental Kinds and Practical Realism

Bruno Mölder

Let me begin with a personal remark about the origins of this contribution.[1] This chapter was prompted by a comment made by Rein Vihalemm at the Estonian Annual Philosophy Conference in 2015, following my presentation on the interpretivist conception of the mind. Interpretivist conception of the mind is the view that interpretation plays a constitutive role in what it takes to have mental states. In that presentation, my aim was to determine whether mental states have any properties that would motivate an interpretivist treatment of them. I settled on the recognition dependence of mental states as the key factor. I argued that for a state to count as mental, it has to be specifiable in mental terms. However, mental states do not wear their label on their sleeve. They do not intrinsically bear their mental specification. Instead, meriting a mental specification is extrinsic—it depends on various factors that are extraneous to the given state, and interpretation is required to pick out the suitable mental specification. I suggested that this kind of recognition dependence sets mental properties apart from natural properties: whether an object has, say, certain physical properties is not constitutively dependent on interpretation. Vihalemm then commented that he does not regard the mental to be special in this respect, for no entity wears its label on its sleeve.

This brief remark turned out to be quite profound. It goes deep into the issue of the relationship between reality and our classifications and has implications for developing a proper conception of natural kinds. This chapter is devoted to making sense of Vihalemm's comment. I will reconstruct the background for the metaphorical idea that no entity wears its label on its sleeve in the context of Vihalemm's own views. This idea can be traced back to Hilary Putnam's rejection of self-identifying objects, but the related influence of Ilkka Niiniluoto's critical scientific realism is also evident in Vihalemm's writings. Thus, this chapter contributes to intellectual history by reconstructing Vihalemm's position through an exploration of both Putnam's and Niiniluoto's positions. I will also consider the upshot of our discussion within the interpretivist framework and provide a response to Vihalemm's comment. Specifically, I aim to uphold interpretivism only concerning the mind, without becoming a global interpretivist, that is, without extending the interpretivist treatment to all kinds of entities.

INTERPRETATION AND MENTAL KINDS

As noted, the version of interpretivism that I considered then, and still consider, worth pursuing is the one that regards the instantiation of mental properties as recognition-dependent. A state is recognized as mental if it is interpreted accordingly. To put it differently, one has those mental properties that are ascribable to one according to the best interpretation.[2]

However, I am not prepared to hold this kind of recognition dependence across the board for all properties. Properly defensible interpretivism should be embedded in a broadly naturalistic framework, acknowledging the existence of entities that are independent of the mind. If the opposite were the case, and the possession of all properties was somehow dependent on interpretation, then one could not be an interpretivist concerning the mental only. One would end up with global interpretivism, in which the whole world is like a text in need of interpretation. However, if the whole world is like a text, then the mental does not differ from the rest. If everything is a matter of interpretation, then nothing is a matter of interpretation! In such a case, the need for a distinctive interpretivist standpoint will disappear, given that interpretivism was initially motivated in part by the need to articulate the sense in which the mental is special when compared to natural properties. In addition, global interpretivism is likely not a coherent position, since it would involve a large interpretative circle. While there are ways of avoiding circularity in the local interpretivist view, the global position would have no recourse to interpretation-independent means for breaking out of the circle.

One way to express the idea that the mental is special, that is, different from natural entities, is to deny that mental kinds are natural kinds. There are different accounts of natural kinds and various ways of conceiving them, but here I take a robust realist approach to natural kinds (cf. Samuels 2009). According to this view, natural kinds are real kinds in nature. They are objective and mind-independent classes of entities. The distinctions between the classes are fairly discrete, and the members of a class are homogeneous. However, this does not yet amount to essentialism, which is the stronger view that natural kinds are determined by their essences.

Even if it is the case that, epistemically, we can only learn about natural kinds through scientific methods, this does not entail that, ontologically, natural kinds are not separate from scientific kinds. On the contrary, it could be said that only those scientific classifications that are useful and productive latch onto natural kinds. As Bird and Tobin (2018) put it, "it is a corollary of scientific realism that when all goes well the classifications and taxonomies employed by science correspond to the real kinds in nature. The existence of these real and independent kinds of things is held to justify our scientific inferences and practices." Scientific practices that can be illuminated by the notion of a natural kind also include drawing inductive generalizations and making discoveries (Samuels 2009, 51). It is natural kinds, as opposed to accidental kinds, that support inductive generalizations. There can be scientific discoveries about natural kinds and their properties by using scientific, empirical methods. These real kinds in nature are neither invented nor constructed by scientists.

On this rather traditional conception, natural kinds are independent of our minds. But what kind of independence is at issue here? The relevant notion of independence here is what Sam Page has called "individuative independence." It is the idea that things or kinds have "boundaries that are totally independent of where we draw the lines" (Page 2006, 327). He also talks about ontological, causal, and structural independence, and natural kinds are independent of us in those senses as well, but those are generally unproblematic and only rarely contested. Thus, an entity is ontologically mind-independent if it would not disappear when we (as minded beings) cease to exist. An entity is causally mind-independent if it was not caused by human minds. An entity is structurally independent of us if its structure is not determined by how we would structure it. Page (2006, 326) notes that most natural entities have some structure and are thus structurally independent of us. His only example of structural dependence is the case where if we were to impose a structure upon an amorphous mass, it would become structurally dependent on our intervention. In this construal, structural dependence is not necessarily implied by individuative dependence, so something can be structurally independent of us even if the way it is individuated depends on us. Page (2006, 328) presents the Goodmanesque example of stars and constellations to illustrate individuative independence. The world is divided into stars with no regard to our individuative contribution, but constellations depend on our activities of identification and conceptualization.[3] However, he points out that constellations are structurally independent of us, presumably because the elements of constellations—namely, stars and their properties—do not depend on us.

Page calls the view that there are natural things or kinds, whose individuation does not depend on us, "individuative realism." As the passage so elegantly encapsulates the core of the position contested by the authors discussed later in this chapter, it is worth quoting in full.

> Individuative Realism is the thesis that the individuation in nature is (metaphysically) real, or in other words, that some things and/or kinds of things in the natural world are not just causally and structurally independent of us, but individuatively independent as well. Just as a turkey is divided up naturally into various joints that are there for the carving, reality itself is divided up naturally into various objects and kinds of objects. (Page 2006, 328)

My point is that we should not be individuative realists about mental kinds. By "mental kinds," I mean the familiar types of mental states that we refer to in everyday life when we talk about beliefs, desires, thoughts, and feelings. It is important to note that the mental is conceived here through folk psychology, the commonsense conceptual framework for making sense of one's own mind and the minds of other people. Thus, the claim that mental kinds are not natural kinds is meant to apply only to mental kinds insofar as they are conceived in folk-psychological terms. It is not intended to be applicable (at least, not without qualification) to the kinds employed in psychological sciences. Those will not be considered in this chapter.[4]

Understood in this sense, mental kinds are not prescribed by the natural world. They result from the ways in which the folk classify and conceptualize the mental realm for various practical purposes. These practical purposes include the prediction and explanation of behavior, social interaction, and regulation of behavior. Mental kinds can be used to form generalizations and make predictions. One could say that mental kinds are practical kinds, to use Zachar's (2000) term—that is, more or less stable patterns that can be reliably identified and that serve practical purposes. As Daniel Dennett (1987, 235) has pointed out, we are skilled in identifying patterns in people's behavior that we interpret in terms of mental states. Practical kinds do not need to carve nature at its joints; they carve it at different places depending on our practical purposes. As noted, the use of mental vocabulary has various practical aims, from self-understanding to manipulating and regulating the actions of other people.

However, these virtues are not yet sufficient to consider mental kinds as natural kinds. Mental kinds are more like constellations than turkeys. Their individuation depends on our folk-psychological practices. To support this claim, I will present four features of mental kinds and classifications that provide reasons why they cannot be regarded as natural kinds. I am not implying that none of these features are exhibited by any natural phenomenon, but when considered together, they make a compelling case. This is not the place to fully defend the view that mental kinds are not natural, so my treatment of these reasons will be relatively brief. For more detail, see Mölder (2019, 181–83), on which the following list is based.

The first reason is the cultural variation and historical contingency of mental classifications (Danziger 1997; Kusch 1999, 241–45). There are diverse folk psychologies that classify the mental realm in various ways, and historically, our folk psychology could have turned out to be different. It can be argued that this variability shows that the individuation of mental kinds depends on cultural contingencies rather than on objective and natural divisions within the mind.

Second, mental kinds are interactive (Hacking 1986). We tend to interact with the descriptions applied to us, and this includes mental ascriptions. Becoming aware of how one's behavior is interpreted may prompt new actions that would not have occurred had the ascription not been made. This phenomenon is well known and aligns with the regulative role of folk psychology. However, a natural kind should not interact with its classification in this way. Hence, there is a reason to assume that mental kinds are not natural.

The third reason is related to the holism of the mental. Ascribing a single mental state is typically not sufficient for predicting behavior. Many other factors should be taken into account, including other ascribable mental states and how they are interrelated. Such holism, however, limits the feasibility of inductive generalizations about mental states. Induction requires uniformity, but holistic mental profiles are always unique, and thus inductive generalization can sometimes lead us astray concerning the matters of the mental. Moreover, everyday mentalistic interpretation usually relies on folk-psychological connections and principles rather than on induction. Thus, while induction is well suited for natural kinds, mental kinds are, again, different.

Finally, while the distinctions between natural kinds are fairly clear and discrete, the boundaries between mental kinds are somewhat vague. The distinctions between mental phenomena are not fixed in the last detail by our folk mental terms. This does not pose issues for the everyday use of folk terms but becomes an obstacle in their use as scientific terms.

These features strongly suggest that mental phenomena do not align with natural kinds. Mental kinds are not individuatively independent and non-interactive kinds with discrete boundaries that would support inductive generalizations without problems.

It is important to note that Rein Vihalemm did not agree with such a traditional picture of natural kinds. He often emphasized that we cannot access the world as it is, independently of our theoretical models, and that those models are necessarily idealized. Vihalemm (2007, 228–29) writes: "[N]atural kinds are not simply 'given' to us by reality, but tell us something about nature only through theories we have constructed, whose idealized models are similar to real systems in specified respects and to specified degrees." At first glance, this may seem like a point about our limited epistemic access to reality, leaving open the possibility that, if we are lucky and successful, our models might resemble the reality out there. However, Vihalemm's point was not merely epistemic. To fully grasp his perspective, we need to consider the distinction he made between two types of cognition in science:

i. φ-scientific cognition, which is constructive, hypothetical, and deductive;
ii. non-φ-scientific cognition, which is classifying, historical, and descriptive.[5]

The prime example of φ-scientific cognition is physics. According to Vihalemm, the physical sciences construct their own objects of study using idealizations and mathematical models. According to his conception, the natural world itself does not constrain natural facts, for natural facts (and, accordingly, also natural kinds) themselves come into play only as constructs in our idealized models. In this context, Vihalemm approvingly cites social constructivists: "The structure, objects, facts, etc. of the natural world are not self-identified by the nature. *In this sense*, the social constructivists are right when they say that 'the natural world has a small, or non-existent, role in the construction of scientific knowledge'" (Collins 1981, 3; Vihalemm 2007, 230). To summarize, according to this view, natural kinds are not part of the natural world but are, instead, constructed within our models of the world.

One cannot help but wonder whether the traditional notion of natural kinds would be more fitting for non-φ-scientific cognition, which, for Vihalemm, is prevalent in fields such as classical biology, which studies natural history, and other nonphysical sciences, such as the humanities.[6] However, this would not easily fit into the traditional conception of natural kinds as objective kinds in nature. For Vihalemm (2007, 230), the objects of non-φ-scientific cognition inevitably possess a "historico-cultural character" and are therefore not "'ready-made' or 'given' by nature itself."

Note that Vihalemm, in the above quote, writes about objects not being "self-identified by nature." The notion of "self-identifying" is significant in this context as it provides a clue to understanding his rejection of entities that wear their label on their sleeve. Entities wearing their label on their sleeve can be understood as self-identifying entities. A philosopher who has explicitly rejected self-identifying entities and whose influence is evident in Vihalemm's writing is Hilary Putnam. His statements on self-identifying objects will be discussed in the next section.

SELF-IDENTIFYING OBJECTS

Hilary Putnam (1981) talks about self-identifying objects in the context of his criticism of metaphysical realism. He construes the position in the following way:

> On this perspective, the world consists of some fixed totality of mind-independent objects. There is exactly one true and complete description of "the way the world is." Truth involves some sort of correspondence relation between words or thought-signs and external things and sets of things. I shall call this perspective the externalist perspective, because its favorite point of view is a God's Eye point of view. (Putnam 1981, 49)

It can be argued that philosophers who support or have supported a realist perspective are not required to adhere to any of these commitments. Putnam's construal of metaphysical realism burdens the view with components that are not necessary for it.[7] A realist does not need to assume that the totality of objects is fixed and that there is only one true description of the world. Neither is a metaphysical realist required to assume a correspondence theory of truth or a God's-eye perspective. That said, we need to proceed from this image of metaphysical realism in order to understand the notion of a self-identifying object.

The perspective opposing this interpretation of metaphysical realism is internal realism, a position Putnam held from the mid-1970s until c. 1990. It basically inverts the view described above. According to internal realism, we can only make sense of the world as divided into objects from within a conceptual scheme. There can be various conceptual schemes, so the world can be divided into objects in more than one way. There is no single complete description of the world. Truth is idealized rational acceptability, not correspondence. Finally, there is no God's-eye point of view; every perspective is laden by the aims and interests of those whose perspective it is. Discussing the objects within a conceptual scheme, Putnam (1981, 52) writes: "'Objects' do not exist independently of conceptual schemes. *We* cut up the world into objects when we introduce one or another scheme of description. Since the objects *and* the signs are alike *internal* to the scheme of description, it is possible to say what matches what."[8] Apart from the conceptual scheme, there can be no privileged relationship between words and objects. To assume that there is something about an object itself such that it merits a certain name or description is to assume that the object is self-identifying.

In *Reason, Truth and History*, Putnam (1981) mentions the term "self-identifying object"[9] only five times on three pages (and using capital letters).[10] Here are the three most telling claims that he makes about them:

I.   Self-identifying objects are "objects that intrinsically correspond to one word or thought-sign rather than another" (Putnam 1981, 51).

II. A metaphysical realist would "say that the word *automatically* covers not just the objects I lassooed, but also the objects which are *of the same kind*—of the same kind *in themselves*. But then the world is, after all, being claimed to contain Self-Identifying Objects, for this is just what it means to say that the *world*, and not thinkers, sorts things into kinds" (Putnam 1981, 53).

III. "[T]he externalist wants to think of the world as consisting of objects that are *at one and the same time* mind-independent and Self-Identifying. This is what one cannot do" (Putnam 1981, 54).

From this list, it is evident that (I) and (II) are different claims. The first one concerns what Putnam dubs "a magical theory of reference," which suggests that objects somehow attract names that are proper to them. The second claim is about the existence of mind-independent kinds in the world, that is, natural kinds. Claim (III) simply states that an object cannot be both self-identifying and mind-independent without providing a reason for this. It is important not to conflate the notion that the world "sorts things into kinds" with the idea that the world also assigns names to these kinds. Objective boundaries between kinds in nature can still exist, even if we are ignorant of them. Let us just assume that our best theory about such boundaries is wrong. It is quite another thing to say that natural kinds intrinsically fall under certain words or labels. Natural kinds are self-individuating, but not self-naming. Naming is what people do.

Perhaps reading (I) as claiming that the objects in the world are self-labeled may not be the most charitable option. Thanks to John Wright (1997, 40–42), there is a more sophisticated reading of Putnam on this issue. In order to illustrate how the world can determine reference, Wright presents two alternative accounts that offer different stories about what determines the referential relationship between a term and an object.

Let us use the example of silver in this context, adapted from Wright's discussion on diamonds. The term "silver" refers to silver. But what makes it refer to silver? According to the first view, the term "silver" denotes any objects that experts consider to be silver under ideal conditions. In this case, the speakers (the experts) have complete control over defining the range of objects that fall under the term "silver." According to the second view, the term "silver" is used for a naturally unified class of things, although the complete properties of that class may not be known. Nonetheless, the term refers to all the individuals belonging to that class. Even if there is a particular specimen of this class that the experts would not recognize as silver due to their imperfect knowledge, this specimen would still belong to the extension of silver because it belongs to the natural class. This particular specimen is an example of a self-identifying object. Putnam would presumably reject the second account, for it entails the assumption that a reference relation can pick out an object independently of the conceptual scheme.

This example links the two sides of self-identifying more closely, as reference is fixed through the affiliation of the kind. Still, (I) reference determined by the world itself, and (II) the world containing kinds independent of us are separate items. Thus, insofar as the notion of a self-identifying object is intended to cover both, the notion remains ambiguous.

It is possible that these two sides of the notion of self-identifying objects led to Vihalemm and me talking past each other during the exchange described at the beginning of this chapter. When I argued, in a rather informal manner, that mental states do not wear their label on their sleeve, I meant that the "labels" are attached to mental states through interpretation and that these labels stem from the conceptual framework known as folk psychology (denial of (I)). I also meant that mental kinds are individuatively dependent on us (denial of (II)). Yet, I did not think that *natural* kinds carry their own label (I); in fact, on that particular occasion, I did not think about the reference relation at all. Nevertheless, I accepted that natural kinds are individuatively independent of us (II). However, Vihalemm may have assumed that by emphasizing that only mental states do not wear their label on their sleeve, I was thereby accepting (I) for natural kinds. By stating that no entity wears its label on its sleeve, Vihalemm may have intended to reject (I). This is one possible interpretation of this exchange, but it is only apparent in hindsight. It is highly plausible that during the actual event, neither of us had a sufficiently articulated understanding of the concept of self-identifying objects, as the notion itself was already ambiguous in Putnam's work. Apart from (I), however, there remains our disagreement on (II) with respect to natural kinds, the idea that natural kinds are individuatively independent of our conceptual schemes. As will become clear later on, despite their differences, Vihalemm would agree with Putnam in rejecting the claim that the natural world "sorts things into kinds," a claim that I am willing to accept.

THE WORLD AND WORLD-VERSIONS

Another significant influence on Vihalemm was Ilkka Niiniluoto's critical scientific realism. Niiniluoto (1999) aims to develop a realist position that would be neither committed to internal nor metaphysical realism, but combines tenets from both. In sum, Niiniluoto's critical scientific realism holds that the objectual organization of the world is relative to the conceptual scheme we use to describe it. It acknowledges that there can be several true descriptions of the world, but conceives truth as correspondence between sentences and the world. The difference between Niiniluoto and Putnam lies in their conception of truth, as well as in the important point that Niiniluoto (1999, 218) does not want to turn the whole world into "human-made construction," a kind of Kantian thing-for-us.

How does Niiniluoto proceed to salvage the real world? He draws a distinction between THE WORLD, which is a mind-independent entity and "world-versions" (a term also used by Goodman 1978). Niiniluoto (1999, 222) regards THE WORLD as "inexhaustible, something that can be described and identified in an unlimited number of ways." World-versions are relative to a conceptual scheme, and each can be seen as giving a "partial description" of THE WORLD. Niiniluoto (1999, 219) notes that "when we 'structure' THE WORLD by our concepts, it is not THE WORLD that changes, but rather our world view. For these reasons, I do not think it is at all incoherent to speak about the mind-independent WORLD." Thanks to science, we could even entertain beliefs about THE WORLD, but those beliefs are fallible.

However, while rejecting the idea that we can only talk about world-versions, Niiniluoto does not conceive of THE WORLD on the model of a Kantian thing-in-itself. He actually thinks that it has its own physical structure of "lawlike flux of causal processes" consisting of tropes or property instances located in space and time (Niiniluoto 1999, 219–21). He rebuffs the notion that before we have applied our conceptual scheme to it, THE WORLD is an unstructured entity, as a "noumenal jam," with reference to Tuomela (1985).[11] Such a jam would get us into trouble if we want to claim that some of our descriptions are true about THE WORLD, for in that case, no sense can be made of the idea that some elements of the jam would correspond to our true statements (Niiniluoto 1999, 217, 225).

If THE WORLD is structurally independent of us, how does this square with the claim that our concepts articulate the world into objects in different ways? Niiniluoto regards the identification and individuation of objects as a human activity that uses a conceptual framework selected for a given purpose. He joins Putnam in rejecting self-identifying objects: "THE WORLD does not contain self-identifying individuals, but can be categorized into objects in several alternative, overlapping ways relative to conceptual schemes" (Niiniluoto 1999, 222). He does not distinguish between identification and individuation and does not allow individuative independence for objects; thus, his critical scientific realism is not individuative realism. Nonetheless, he acknowledges that "the possibility of the identification of a physical thing, like a dinosaur or a chair, is indeed based on its mind-independent features (location in space and time, causal continuity, qualities)" (Niiniluoto 1999, 221). However, it still remains unclear if this conception can make sense of the (useful) idea that some of our identifications and classifications in our world-versions correspond better to THE WORLD than others. After all, the objective structure of the world as a "lawlike flux" seems rather different from our alternative conceptualizations tailored to human needs and purposes.

PRACTICAL REALISM AND SELF-IDENTIFYING OBJECTS

Let us return to Rein Vihalemm. His practical realism, rooted in the ideas of Karl Marx and Thomas Kuhn, also aims to present a third option alongside internal and metaphysical (or, as he calls it, "standard scientific") realism. He rejects metaphysical realism for presuming the God's-eye viewpoint and ignoring human practice. Regarding Putnam's internal realism, Vihalemm (2011, 49) believes that it inevitably falls into conceptual idealism and, therefore, cannot be considered a realist position. Vihalemm (2012, 18) interprets Niiniluoto's critical scientific realism as a version of practical realism, but notes that it pays too little attention to science as a practical activity, and he does not share Niiniluoto's view on truth as correspondence.

Vihalemm (2011, 48) presents practical realism in the form of five theses.[12] Note that these concern explicitly science and scientific practice, not other kinds of human activities or mind-world relationship in general, but it is plausible to assume it was intended rather broadly. The first thesis rejects metaphysical realism: "science does not represent the world 'as it really is' from a god's eye point of view," but then the second thesis blocks the recoil to internal realism: "the world is not accessible independently of theories—or . . . practices," but this "does not mean that Putnam's internal realism (or social constructivism) is acceptable." The next three theses concern science as practice. The third thesis states that science is mainly "a practical activity whose main form is scientific experiment which in its turn takes place in the real world." The fourth thesis stresses the normativity of science: it is "also a social-historical activity . . . that includes a normative aspect . . . and that means . . . that the world as it is actually accessible to science is not free from norms either." Fifth, he stresses that practical realism "is certainly realism as it claims that what is 'given' in the form of scientific practice is an aspect of the real world." In sum, to use Niiniluoto's terminology, the basic point of practical realism is that we can access THE WORLD only through our scientific practices by which our practical world-versions become imbued with norms.

Vihalemm has also remarked on the structure of the world and self-identifying objects. In this regard, it is somewhat difficult to discern his own views, as they are often presented through the exposition and endorsement of Niiniluoto's position.

It should be acknowledged that the scientific account of the world is mediated by our practical and theoretical activity, together with our aims and values, which means that our descriptions of the world, our "world-versions," are always relative to us. This does not imply, however, that the world itself (we can call it THE WORLD) is relative to us in the sense that our "world-versions" cannot be versions of THE WORLD (see Niiniluoto 1999, 218–226). Our scientific "world-versions" . . . still do tell us something about THE WORLD, as do the theories we have constructed, which, in their theoretical models, contain experimentally substantiated idealisations, since . . . theoretical models are similar to real systems in specified respects and to specified degrees. (Vihalemm 2012, 17–18; see also Vihalemm 2003, 66–67)

It is only through practice that the objective world can really exist for humans. . . . We are not "world makers." The world, however, does not consist of self-identifying objects; objects are identifiable—in principle, in a potentially infinite number of ways (in this sense they are inexhaustible, having innumerable aspects and connections with the rest of the world)—through practice. (Vihalemm 2012, 10)

On the face of it, Vihalemm's position on THE WORLD is very similar to Niiniluoto's, but with the qualification that we represent it through our practices. Following Niiniluoto, who in turn followed Putnam, Vihalemm claims that there are no self-identifying objects in THE WORLD. However, there is a difference between Niiniluoto's and Vihalemm's understanding of THE WORLD. Vihalemm claims:

I take THE WORLD to be unidentified objective reality or matter, objective in the absolute sense, i.e., independent from anyone's mind or consciousness; this absolute objectivity of its existence is its only defining characteristic, it is "matter as such." It was the "thing-in-itself " for Kant; however, for practical realists or materialists it is not ungraspable, but identifiable in its concrete forms of existence through practice, being itself a concrete way of objective existence. (Vihalemm 2012, 19)

As mentioned above, Niiniluoto does not equate the world with the Kantian thing-in-itself; he takes it to consist of causally related tropes. Note that although Vihalemm allows that we can access and identify THE WORLD only through practice, he still imputes to it one property—"absolute objectivity." If this property is supposed to be a practice-independent property, then it is not consistent with his assumption that the identification of the properties of THE WORLD *only* occurs through practice. In addition, there is a concern that if THE WORLD is an "unidentified objective reality," it seems very much like the "noumenal jam" that Niiniluoto wanted to avoid. It would lack intrinsic structure, as all structure is imposed on it through our models. In contrast, Niiniluoto allowed THE WORLD to be structurally independent of us. It is only individuatively dependent on us. Vihalemm acknowledges that models can be similar to reality, but it is not clear how this can be redeemed in his view—how can we determine which models are similar to THE WORLD and which are not, if the God's-eye point of view is not possible?

As THE WORLD in itself is an unidentified totality, natural kinds belong firmly to the side of world-versions, being denizens of our models: "our 'world versions' (including natural kinds identified by us), but not the world itself, are relative to us" (Vihalemm 2003, 67).[13]

I believe this is sufficient to clarify the intellectual background for Vihalemm's comment that it is not just mental entities that wear their label on their sleeves. It expresses his rejection of the existence of self-identifying objects and his conception of natural kinds as being relative to the models we have constructed through scientific practice.

INSTEAD OF CONCLUSION: BACK TO THE REAL WORLD

Having clarified Vihalemm's position, I conclude this chapter by outlining a response that is consistent with my commitment to an interpretivist approach to the mind. As far as I can see, there are two options regarding practical realism and interpretivism.

The first option is to accept practical realism but still try to maintain the distinction between mental kinds and natural kinds. While all kinds require human intervention to attach labels to their sleeve, the mental kinds require more substantial sewing. Mental kinds would be more interpretation-dependent than natural kinds, perhaps for the reasons presented in section "Interpretation and Mental Kinds," provided that those do not apply to natural kinds.

However, I believe that this response is too weak, as it makes the distinction between mental and natural kinds rather ambiguous and indefinite. From a practical realist perspective, it becomes challenging, if not in principle impossible, to differentiate between the projected and reflected elements in our conceptions. Yet, disentangling the contribution of THE WORLD from the contribution of our practices is crucial, particularly for those approaches that aim to conceive some parts of our models or conceptions as projections (such as mental properties in the case of interpretivism).

When Vihalemm stated that natural kinds, *as identified by us*, are relative to us, this is trivial if one considers identification to be a human activity. An individuative realist who assumes that there are natural kinds independent of our identifications could even agree with this claim ("Sure, natural kinds for us are relative to us!"), but would maintain that this does not imply that natural kinds as independent joints in nature are relative to us.

We need independent natural kinds to make sense of the crucial role of the natural world in forming our models: THE WORLD provides hints of its existence and gives clues about its properties by either cooperating with us or resisting us. This sort of contribution from THE WORLD makes it possible to regard some descriptions of the world as better than others—and they are objectively better, not just better from within a conceptual scheme or practice. Our true descriptions require objective distinctions in THE WORLD.[14] I agree with David Lewis (1984, 228), when he says (in response to Putnam) that "the realism that recognises a nontrivial enterprise of discovering the truth about the world needs the traditional realism that recognises objective sameness and difference, joints in the world, discriminatory classifications not of our own making."

Of course, we are bound by our perspectives, but we can still make the metaphysical assumption that our best scientific classifications are backed up by the natural kinds that are out there in the objective world. Moreover, it could be said that due to the fact that our brains have evolved in causal response to the objective kinds, we are able to attain knowledge about them (cf. Blackburn 1999, 268).

That brings me to the second and preferred option. It would combine individuative realism about the natural world with interpretivism about the mental. This means rejecting the practical realist view that we construct and identify natural kinds only through scientific practice and that the world in itself is just a "noumenal jam" waiting to be identified. Natural kinds are out there individuatively independent of us. What we can discover about them is determined by the nature of THE WORLD. If we classify them incorrectly, our errors have consequences, and THE WORLD itself contributes to highlighting these errors. THE WORLD "strikes back" when we fail to classify its kinds correctly, and this requires individuative rather than just structural independence. As an interpretivist, I do not hold the same view on mental kinds (as individuated in folk-psychological terms). The difference can be formulated in the following way: THE WORLD would still divide into natural kinds even if there were no scientific practice, but there cannot be mental kinds without the practice of folk psychology. This approach allows respecting the distinction between mental and natural kinds.

NOTES

1. My research on this chapter has been supported by the European Union, European Regional Development Fund (Centre of Excellence in Estonian Studies, TK145).

2. For details of such an interpretivist position, see Mölder (2010).

3. Goodman (1996, 145) himself held an opposing view: "We make a star as we make a constellation, by putting its parts together and marking off its boundaries."

4. The relationship between folk psychology and the classifications of scientific psychology is a substantial topic that requires a separate study. For some of my previous attempts to tackle this issue, see Mölder (2016, 2–13; 2017, 58–61).

5. Vihalemm has elaborated on this distinction in various places, compare, for instance, Vihalemm (2007, 2016).

6. See also the remark in Vihalemm (2003, 61) that "[t]he problem of natural kinds stands differently in natural history and φ-*science*, and it is a difference of principle."

7. See Sankey (2018), who makes a convincing case for this claim.

8. This quote contains what is known as the "cookie-cutter" metaphor. Later, Putnam (1987, 36) came to criticize it for the metaphor presumes that it is still possible to speak about one substance—the dough—that can be divided in various ways, but for an internal realist, everything, including the dough, must be relative to a conceptual scheme.

9. Putnam credits Wiggins (1980) for the origin of the term, but Wiggins writes (somewhat cryptically) about "self-differentiating," not self-identifying objects: "the realist myth of the *self-differentiating object* (the object which announces itself as the very object it is to any mind, however passive and of whatever orientation)" (Wiggins 1980, 139).

10. The term occurs also in *Realism and Reason*, in connection with the reference relationship: "the idea that . . . *nature itself* determines what our words stand for—is totally unintelligible. At bottom, to think that a sign-relation is *built into nature* is to revert to medieval essentialism, to the idea that there are 'self-identifying objects' and 'species' out there" (Putnam 1983, xii).

11. I have to confess that while there is some discussion of noumena in Tuomela (1985), I could not find any mention of "noumenal jam" in that book.

12. See also Vihalemm (2005, 180–81).

13. The additions in square brackets Vihalemm (2012, 18) made to the following Niiniluoto's quote are also telling in this regard: "If we use the cookie-cutter metaphor, we can say, 'A cake [THE WORLD—R.V.] can be sliced into pieces in a potentially infinite number of ways, and the resulting slices [say, natural kinds and laws of nature identified by us—R.V.] are human constructions made out of the parts [unidentified (complex, inexhaustible) objects, their properties and relations—R.V.] of the cake'" (Niiniluoto 1999, 222).

14. One might wonder if a similar line of reasoning would also support realism about the mental, given that it is the world that makes some mental descriptions better than others. However, this line of reasoning is blocked when it comes to the mental "realm." This is because the standards for favoring certain mental descriptions over others are internal to the interpretation of mentalistic discourse, rather than external, as in the case of natural kinds. These standards are not determined by objective, substantial mental or neural facts; instead, choosing one mental description over another involves interpretation. (I am grateful to an anonymous reviewer for pressing this point.)

REFERENCES

Bird, Alexander, and Emma Tobin. 2023. "Natural Kinds." In *The Stanford Encyclopedia of Philosophy* (Spring 2023 Edition), edited by Edward N. Zalta and Uri Nodelman. Accessed October 2023, https://plato.stanford.edu/archives/spr2023/entries/natural-kinds.

Blackburn, Simon. 1999. *Think: A Compelling Introduction to Philosophy*. Oxford: Oxford University Press.

Collins, Harry M. 1981. "Stages in the Empirical Programme of Relativism." *Social Studies of Science* 11 (1): 3–10. https://doi.org/10.1177/030631278101100101.

Danziger, Kurt. 1997. *Naming the Mind: How Psychology Found Its Language*. London: Sage. https://doi.org/10.4135/9781446221815.

Dennett, Daniel C. 1987. *The Intentional Stance*. Cambridge, MA: MIT Press.

Goodman, Nelson. 1978. *Ways of Worldmaking*. Indianapolis, IN: Hackett. https://doi.org/10.5040/9781350928558.

Goodman, Nelson. 1996. "On Starmaking." In *Starmaking: Realism, Anti-realism, and Irrealism*, edited by Peter J. McCormick, 143–47. Cambridge, MA: MIT Press.

Hacking, Ian. 1986. "Making Up People." In *Reconstructing Individualism*, edited by Thomas C. Heller, Morton Sosna, and David E. Wellbery, 222–36. Stanford, CA: Stanford University Press.

Kusch, Martin. 1999. *Psychological Knowledge: A Social History and Philosophy*. London: Routledge.

Lewis, David. 1984. "Putnam's Paradox." *Australasian Journal of Philosophy* 62 (3): 221–36. https://doi.org/10.1080/00048408412340013.

Mölder, Bruno. 2010. *Mind Ascribed: An Elaboration and Defence of Interpretivism*. Amsterdam: John Benjamins. https://doi.org/10.1075/aicr.80.

Mölder, Bruno. 2016. "Mind and Folk Psychology: A Partial Introduction." *Studia Philosophica Estonica* 9 (1): 1–21. https://doi.org/10.12697/spe.2016.9.1.01.

Mölder, Bruno. 2017. "Mind Re-ascribed." *Studia Philosophica Estonica* 10 (2): 55–104.

Mölder, Bruno. 2019. "Vaimuseisundid pole loomulikud liigid." [in Estonian] In *Humanitaarteadused ja kunstid 100-aastases rahvusülikoolis*, edited by Riho Altnurme, 173–85. Tartu: Tartu University Press.

Niiniluoto, Ilkka. 1999. *Critical Scientific Realism*. Oxford: Oxford University Press.

Page, Sam. 2006. "Mind-Independence Disambiguated: Separating the Meat from the Straw in the Realism/Anti-realism Debate." *Ratio* 19 (3): 321–35. https://doi.org/10.1111/j.1467-9329.2006.00330.x.

Putnam, Hilary. 1981. *Reason, Truth and History*. Cambridge: Cambridge University Press. https://doi.org/10.1017/CBO9780511625398.

Putnam, Hilary. 1983. *Realism and Reason: Philosophical Papers, Volume 3*. Cambridge: Cambridge University Press. https://doi.org/10.1017/CBO9780511625275.

Putnam, Hilary. 1987. *The Many Faces of Realism*. La Salle, IL: Open Court.

Samuels, Richard. 2009. "Delusion as a Natural Kind." In *Psychiatry as Cognitive Neuroscience: Philosophical Perspectives*, edited by Matthew Broome and Lisa Bortolotti, 49–80. Oxford: Oxford University Press. https://doi.org/10.1093/med /9780199238033.003.0004.

Sankey, Howard. 2018. "Putnam's Internal Realism in Retrospect." *ARIF: Análisis. Revista de Investigación Filosófica* 5 (1): 27–50. https://doi.org/10.26754/ojs_arif /a.rif.201812921.

Tuomela, Raimo. 1985. *Science, Action, and Reality.* Dordrecht: Reidel. https://doi. org/10.1007/978-94-009-5446-5.

Vihalemm, Rein. 2003. "Natural Kinds, Explanation, and Essentialism in Chemistry." *Annals of the New York Academy of Sciences* 988: 59–70. https://doi.org/10.1111 /j.1749-6632.2003.tb06085.x.

Vihalemm, Rein. 2005. "Chemistry and a Theoretical Model of Science: On the Occasion of a Recent Debate with the Christies." *Foundations of Chemistry* 7: 171–82. https://doi.org/10.1007/s10698-005-0959-y.

Vihalemm, Rein. 2007. "Philosophy of Chemistry and the Image of Science." *Foundations of Science* 12 (3): 223–34. https://doi.org/10.1007/s10699-006-9105-0.

Vihalemm, Rein. 2011. "Towards a Practical Realist Philosophy of Science." *Baltic Journal of European Studies* 1 (1(9)): 46–60.

Vihalemm, Rein. 2012. "Practical Realism: Against Standard Scientific Realism and Anti-Realism." *Studia Philosophica Estonica* 5 (2): 7–22. https://doi.org/10.12697 /spe.2012.5.2.02.

Vihalemm, Rein. 2016. "Science, φ-Science, and the Dual Character of Chemistry." In *Essays in Philosophy of Chemistry*, edited by Eric Scerri and Grant Fisher, 352–79. New York: Oxford University Press. https://doi.org/10.1093/oso /9780190494599.003.0024.

Wiggins, David. 1980. *Sameness and Substance.* Oxford: Blackwell.

Wright, John. 1997. *Realism and Explanatory Priority.* Dordrecht: Springer. https:// doi.org/10.1007/978-94-017-2844-7.

Zachar, Peter. 2000. "Psychiatric Disorders are Not Natural Kinds." *Philosophy, Psychiatry, & Psychology* 7 (3): 167–82.

Part III

SPECIAL SCIENCES

Chapter 7

Practical Realism and the Philosophy of the Humanities

Sami Pihlström

The realism debate traditionally focuses on the general philosophy of science and the philosophy of the natural sciences.[1] It is in this context that Rein Vihalemm made his lasting contributions to understanding the realism issue as primarily *practical*, emphasizing that our operations of inquiry taking place in the material world, including systematic and controlled experimentation in particular, open us toward a cognizable reality. Scientific knowledge cannot be disentangled from such practical action—or what the classical pragmatists (i.e., Charles S. Peirce, William James, and John Dewey) would have called habits of action. Vihalemm's "practical realism" is partly grounded in Karl Marx's account of practice, but his basic position can, in my view, be plausibly cashed out in pragmatist terms, too. In general, it can be seen as one important manifestation of what has become known as the orientation in contemporary philosophy of science insisting on the significance of *scientific practices*—in contrast to primarily investigating the logical structure of scientific theories. In this respect, Vihalemm's approach resembles, for example, Thomas Kuhn's ([1962] 1970) well-known theory of paradigms (which obviously cannot be regarded as realistic in any standard sense, though) and, among more recent examples, Joseph Rouse's (2002) and Hasok Chang's (2022) views on realism and practice.

It is important to observe that the realism debate is not restricted to the philosophy of the natural sciences, even though the most widely discussed examples—such as the question concerning the mind- and theory-independent reality of unobservable theoretical entities like electrons and black holes—are typically drawn from the sciences. Analogous questions about the theory-independent (vs. theory-dependent) existence of elements of human social and cultural reality may be raised within the human sciences, though it would hardly make sense to claim the research objects of the humanities

163

to be "mind-independent," because they are in most cases literally created by humans. The realist about the humanities may claim, analogously to scientific realism, that while the world as investigated by the humanities is humanly constructed and receives any ontological categorization it may be claimed to have only in rich theoretical frameworks or traditions of scholarly interpretation, there is a sense in which the humanistic scholar is seeking the truth about the way(s) the world is in their field of study. The reality described, explained, and interpreted in the humanities need not be *specifically* ontologically dependent on the theorist's theory, or the interpreter's interpretation, although it is *in general* dependent on interpretive practices possibly utilizing complex theoretical discourses and approaches.[2]

While philosophers of social science have contributed to scientific realism both within general philosophy of social science and special fields such as the philosophy of economics,[3] the distinctive perspective of the humanities is traditionally not very strongly represented in these discussions. This chapter will therefore introduce the realism issue in what we may call the philosophy of the humanities and will defend a *pragmatist* approach to the debate, briefly articulating a form of *pragmatic realism* especially regarding the ontology of the theoretical postulations of humanistic inquiry (which is something I have more comprehensively defended in Pihlström [2022]). In a more detailed investigation, it would have to be considered whether different fields within the humanities, such as historiography, literary theory, or theology and religious studies, should be treated in different ways with regard to the problem of realism. Here I will merely employ a simple example drawn from religious studies in order to highlight the pragmatically reflexive character of the realism issue. The relation between pragmatic realism (emerging from the pragmatist tradition but also, as I will explain, from Kantian transcendental philosophy) and Vihalemm's practical realism will be explored in this context, taking for granted that any pragmatic or practical realism—regarding the sciences as much as the humanities—will have to be cautious in making any claims about an ontologically "independent" reality, as any reality we can meaningfully engage with is always already a reality conceptualized through our practices of inquiry. In this regard, there is no essential difference between the sciences and the humanities.

PRACTICAL REALISM

In earlier work spanning over several decades, I have defended a version of pragmatic realism seeking to integrate pragmatist philosophy of science and inquiry with an acknowledgment of ordinary (non-metaphysical) realism about the world that we live in and seek to know and understand more deeply through our inquiries which, reflexively, continuously refine and ameliorate their own methods and assumptions.[4] Before explaining what this form of realism comes down to in the philosophy of the humanities, let me reflect in broad strokes on the relation between Vihalemm's "practical realism" and what I am calling "pragmatic realism." This is how Vihalemm characterizes his practical realism:

> To speak about the world outside practice means to speak about something indefinable or illusory. It is only through practice that the objective world can really exist for humans. Therefore, knowledge must be regarded as the process of understanding how the world becomes defined in practice. One should say that science as practice is a way that we are engaged with the world and that allows the world to show how it can be identified in its own possible "versions." We are not "world makers." The world, however, does not consist of self-identifying objects; objects are identifiable—in principle, in a potentially infinite number of ways (in this sense they are inexhaustible, having innumerable aspects and connections with the rest of the world)—through practice. And practice is, in short, human activity as a social-historical, critically purposeful-normative, constructive, material interference with nature and society producing and reproducing the human world—culture—in nature. (Vihalemm 2012, 10)

The world as conceptualized and cognized by science is thus a world conceptualizable and categorizable only through human practices. Denying not only the actuality but even the possibility of "self-identifying objects," Vihalemm emphasizes that it is from within and in relation to our practices that any objects we take to be real are identified, or indeed *identifiable*, as the objects they are. This, in my view, comes very close to the pragmatism defended by philosophers like James and Dewey—or the more recent pragmatism developed by Hilary Putnam (e.g., 1981, 1990), whose one-time "internal realism" was an influential (though, according to Putnam himself, failed) attempt to bridge the gap between realism and pragmatism in the philosophy of science (cf. Putnam 2016).

In some of his insightful papers on these topics, Vihalemm (2011, 2012) challenged me to compare pragmatic realism with his practical realism. I do not want to quarrel about words, and I warmly welcome his proposal, for example, to examine more closely Marx's concept of practice in relation to the realism issue and to view it as analogous to pragmatist notions. However, I still maintain that my own pragmatic realism, which, unlike Vihalemm's practical realism, incorporates (albeit in a pragmatically rearticulated form) the crucial idea of the practice-laden construction of the empirical (knowable, experienceable) world adapted from Kantian transcendental idealism, offers a plausible approach to the realism debate. Indeed, Vihalemm's tendency to soften the boundary between the concepts of construction and identification (of objects) in the context of scientific practices supports this Kantian-inspired account, which integrates empirical realism with what we may call transcendental pragmatism. Insofar as by identifying scientific objects we in a sense not just discover but constitute them, or rather their possibility as the kind of objects they are, within our practices of inquiry (including both scientific practices and the practices of inquiry within social-scientific and humanistic disciplines), those objects—and practices—clearly cannot be accounted for in terms of metaphysical realism (to employ Putnam's vocabulary). No objects are mind- and theory-independently real from an absolute God's-eye view, but their reality (viz., any reality they can be meaningfully claimed by us to enjoy) is practice-embedded, or practice-internal, and hence humanly perspectival.[5] The objects of inquiry thus do not possess any "ready-made" identity prior to our practices of inquiry—and here I obviously agree with Vihalemm. Rather, as Dewey (e.g., [1929] 1960), among other pragmatists, insisted in his critique of the "spectator theory of knowledge," scientific inquiry does not aim at, nor can it be based on, passive contemplation of eternal unchanging truths about what is really there independently of inquiry but must be understood as a critically self-corrective practice whose objects emerge from the processes of inquiry themselves instead of being "there" to be discovered with a preexisting ontological identity prior to inquiry.

In this respect, I am also pleased to join Vihalemm's polite criticism of Ilkka Niiniluoto's (1999) highly sophisticated yet considerably stronger form of realism, "critical scientific realism," which postulates a realistic ontology of a completely mind- and theory-independent world consisting of (at least in principle) *unidentified* objects until conceptually categorized (i.e., objects existing, with ontological identity, prior to and independently of inquiry) and incorporates a version of correspondence truth—a traditional Aristotelian idea made precise by utilizing Alfred Tarski's model-theoretic truth-definition. As neither the practical realist nor the pragmatic realist is willing to go that far in their endorsements of realism, Vihalemm's position and mine are, I believe, more united than divided here, no matter whether the view is labeled pragmatic or practical realism. We share the firm rejection of any "absolute," ontologically pre-categorized way of the world. On the other hand, Vihalemm (2012, 17–18) agrees with Niiniluoto's rejection of Putnam's internal realism (which, precisely due to its Kantian dimensions, is for these realists not a form of realism at all), while also maintaining that practical realism can accommodate a form of semantic realism by endorsing a deflationary conception of truth (instead of the correspondence theory). It is clear that my own pragmatic realist approach comes closer to Putnam's form(s) of realism and pragmatism, though I would also urge (and have urged) the Putnamian pragmatic realist to formulate their position in transcendental terms, which Putnam himself was always reluctant to do, presumably fearing the same threat of idealism that both Niiniluoto and Vihalemm have wanted to avoid—without, in my view, appropriately distinguishing between transcendental and empirical idealism.[6]

While Niiniluoto, for Vihalemm's taste, is one of those traditional realists whose conception of science remains too abstract, neglecting scientific practice, Vihalemm insists that scientific realism as developed in the philosophy of science today must take seriously the inherent practicality of science. This view, shared by many realists and non-realists alike, entails taking seriously the further idea, strongly emphasized by Vihalemm and other representatives of practice-based philosophy of science, that scientific knowledge is produced and scientific objects and facts identified (or even constructed) within a rich *plurality* of practices of inquiry. Pragmatism, again, can very well accommodate this pluralism not only in a conceptual or methodological sense but even in an ontological sense. Accordingly, we also need to understand the intertwinement of our diverse scientific and scholarly practices, as well as the *interdisciplinary* character of those practices of inquiry, whenever relevant. Reality itself is "plural" in the sense of not being reducible to any overarching privileged practice of conceptualization.

A practical, localized, and contextualized approach to the realism issue hence encourages us to relativize the opposition between realism and antirealism in its different dimensions to its practical contexts. Our interpretations of scientific theories and their ontological postulations—that is, theoretical concepts and/or entities—must be contextualized not only within areas of discourse but also within practice-laden, and often interdisciplinary, inquiries embodied in our habitual actions taking place in the material world. Moreover, not only is the identity of the postulated entities a contextual, practice-embedded matter (as Vihalemm argues); at the meta-level, the distinction between the realistic context- or practice-independence of such postulations, on the one hand, and their dependence on contexts or practices, on the other, must itself be contextualized into practices of philosophical inquiry (cf. Pihlström 2020, chs. 1–2). This reflexivity could in principle continue indefinitely.

Now, I am not entirely convinced that Vihalemm's practical realism is fully equipped to account for such endlessly reflexive practice-embedded contextualizing. This is because there is a sense in which it still views scientific practices "from above," or from an allegedly fully objective vantage point beyond those practices themselves. Let me elaborate on this.

One might suggest that the following problem arises.[7] Can we simultaneously interpret (say) a theory *T* realistically within inquiry (practice) *I* and nevertheless interpret it antirealistically within another, possibly overlapping (context of) inquiry *I'*, assuming that *T* is in some way employed—for instance, either presupposed or critically tested—within both practical contexts, I and I'?[8] If inquirers are allowed, or required, to "mix" their disciplinary identities and practices when engaging in interdisciplinary inquiries, such problems may have to be faced even in a relatively practical sense. Broadening our scope from the philosophy of science to cover humanistic, and possibly ethical and other value inquiries, we might ask whether it is possible to understand, say, moral values and norms realistically when engaging in moral philosophy (including the ethics of science) and, at the same time and/or by the same inquirer, antirealistically (i.e., as reducible to, e.g., natural interests of survival explainable with reference to natural selection, and thus ontologically speaking ultimately "something else") when engaging in an interdisciplinary inquiry into the evolutionary origins of moral behavior incorporating biological, psychological, social-scientific, and other practices. Can one and the same person—or a group of investigators, or a research project, or a Peircean idealized community of rational inquirers (whatever the "subject" of inquiry may be)—operate in terms of both interpretations, locally turning from a realist to an antirealist interpretation of their theories (and practices), and back again, depending on the practices of inquiry they are at a given time engaging in?

In the kind of interdisciplinary situation we are imagining here, in which T is in the "common area" of two (or more) overlapping inquiries constituting an interdisciplinary set of practices, we presumably cannot interpret T either realistically or antirealistically from a global philosophical perspective. A practice-based philosophy of science and inquiry must insist that we always have to interpret it contextually, that is, in the context of, say, I or I' (or some other inquiry or practice). There is never a metaphysically speaking independent object of inquiry beyond all such contexts, as our natural habits of action—as the classical pragmatists already saw—are involved in the ontological identity of the objects we engage with. This may also require a reconsideration of our own identities as inquiring subjects. It may be a deeply perspectival and context-dependent matter whether we are committed to a realistic or an antirealistic account of T and its theoretical postulations. This contextuality of not only the *objects* but also the *subject* of inquiry is something that a "practical realism" in the philosophy of science ought to recognize, along with taking ontologically seriously the practice-embeddedness of the identities of any objects of inquiry, precisely because pragmatist philosophy of inquiry cannot start from any traditional dualism between the subject and the object anyway. The kind of transcendentally grounded pragmatic realism I have formulated on other occasions (e.g., Pihlström 2020, 2021, 2022, 2023b) may, I submit, be better equipped to explore this contextuality than Vihalemm's (non-transcendental) practical realism is. This is because such a pragmatic realism emerges from the Kantian-inspired idea that it is through our (in a broad and pragmatically reinterpreted sense) "subjective" practice-embedded perspectives on the world that the world gets the (or any) ontological shape(s) it does.[9] Vihalemm's own remarks on the identifiability of objects through practices can, I submit, be reinterpreted along these Kantian lines, although he was never himself happy about such interpretations.

Given Vihalemm's (philosophical, not political) Marxism, I am afraid his practical realism and my quasi-Kantian pragmatic realism are in the end obliged to view the relation between the subject and object differently. He writes:

> In Marxist philosophy, the subject and its practical activity, becoming a legitimate part of material reality (objective reality), also have objective characteristics. Thus, the subject is included in material reality as a specific component and no longer has consciousness as its only constituent property. The impact of practice on reality is brought about not from "outside" but from "inside" the latter. This is the impact of one form of objective reality on another—the impact of reality "in the form of activity" on reality "in the form of an object." (Vihalemm 2012, 13)

From the pragmatic-transcendental standpoint, this Marxist understanding of practice still takes a sideways-on perspective on the subject's activity within the practice(s) they engage in. The practices themselves are viewed from a God's-eye view, as it were. If this is the case, then no full-blown contextualizing (or recognition of the inevitability of such contextualizations) from within the practices themselves can take place.

It may be suggested, furthermore, that the practical realist about science must extend their realism to cover the social and cultural features of the practices they view scientific theorization as dependent on. This is a concrete manifestation of the kind of reflexivity of practices I have loosely invoked above. That is, the practical realist is not only a realist about, say, the practice-laden postulations of physics and chemistry but also about the features of the practices enabling those postulations to be made. Such a practical realism about human practices themselves, particularly about the practices of inquiry we engage in, may have to accommodate a realistic theory of value and normativity, for example (though again a realism pragmatically contextualized). Without practices guided by values and purposes, there can be no scientific experimentation and inquiry at all—no science in the sense in which we know it. This is one obvious reason why the practical realist philosophy of science Vihalemm has so powerfully offered us must be extrapolated to the philosophical study of the social and the human sciences, seeking to interpret the practices we engage in in an analogously realistic manner. But our realism in those areas must, again, itself be practically grounded. Practical realism about human practices, a realism needed for a realistic account of science to be possible, must itself be a practical realism (or, in my terms, pragmatic realism).

Now, this reflexivity is something that a *pragmatist*, as distinguished from Vihalemm's (Marxist) practical conception of inquiry, can in my view best accommodate. For the pragmatist, and thus for the pragmatic realist, there is no way of viewing the practices enabling our scientific realism from a side-ways-on perspective. We are always already within those practices, engaging in them and committed to their norms that guide our engagements, and thus interpreting our natural and social world through them, or (better) as entangled with them. Moreover, we always already operate within a multiplicity of practices. We cannot simply leave our commitment to the norms of inquiry *I* behind when considering the compatibility of our postulations within *I* to those within *I'* in an interdisciplinary context (or the interpretations of a shared theory *T* within these domains). Rather, our practically realist understanding of the objects of inquiry as constituted within our normatively governed practices is only possible for us from within such normatively governed practices themselves. What it is to be committed to norms of inquiry ought to be investigated from a standpoint within those norms, fully acknowledging that this investigation cannot be adequately conducted from any imagined nonnormative Archimedean point beyond those normative practices. Here, a reflexively pragmatist philosophy of science enabling a pragmatic scientific realism conscious of its roots in Kantian critical philosophy gains the upper hand.

AN EXAMPLE FROM (THE PHILOSOPHY OF) RELIGIOUS STUDIES

In order to emphasize the plurality of the practices of inquiry we need to engage in for a truly practical scientific (or humanistic) realism to be as much as possible for us (from within those practices), let us imagine real-life practices of inquiry into something that can receive both natural-scientific and humanistic interpretations. A scholar in religious studies may be interested in the ways in which religious believers participate in rituals, perceive and use "religious artefacts" carrying certain symbolic religious meanings, and interpret religious texts (cf. Kalmykova 2024). Such a scholar may, for example, study the religious meanings found or even "perceived" in a bottle of (allegedly) sacred water taken from a fountain that is believed to have spiritual properties. The scholar knows, of course, that the "holy" water is, scientifically speaking, just water. The scholar may, moreover, be a full-blown scientific realist believing in the ontologically interpreted theory-independence of the chemical properties of water (which, he or she might further believe, are ultimately reducible to ontologically fundamental physical properties); this realism may, along Vihalemm's lines, be grounded in the practices of scientific inquiry. At the same time, the scholar might interpret the equally practice-based "theories" of the religious believers one studies along antirealistic lines. The claim that the water possesses spiritual properties (e.g., a divine presence of some sort, or mysterious healing powers) is not, according to the scholar's interpretation of what is going on in the religious believers' practices, to be taken as realistically seriously, ontologically speaking, as the interpretation of either the scholar's own or the religious practitioners' scientific theory about water. Indeed, the practitioners under investigation may very well share the scientific conception of what water is and may firmly believe that scientifically speaking the doxastically "holy" water to which they attach symbolic properties is "just water." As Elena Kalmykova (2024) notes in her insightful book on religious beliefs and practices, such believers may even know that the water they use for their religious purposes comes from the tap instead of coming from any sacred fountain. Yet, even then they may treat the water as a religious artifact playing a role in their rituals, bridging the gap between their perceptions and the transcendent they (in some sense) believe to be real.

Note that this is *not* simply to claim that because the water the religious practitioners are dealing with is "just water," their "theory" about it is false. That would be far too simple and would commit the easy error of confusing religious beliefs with scientific beliefs—an error any pragmatist should avoid (cf., e.g., Pihlström 2020). The religious studies scholar we are imagining here stands at the intersection of scientific and humanistic perspectives on the reality he or she is investigating. From within the practices of inquiry in religious studies, the scholar is *not*, of course, inquiring into the chemical structure of water, even the water that the religious people studied are dealing with—though it may be relevant for the scholar's interpretation of them to know that they also believe the ritualistically sacred water to be in a scientific sense, or under a scientific description, "just water."[10] On the contrary, the religious studies scholar is studying those people and their beliefs and practices—not, of course, water. Within the practice of inquiry he or she is engaging in, the chemistry of water is not at issue; the scholar can very well just presuppose a physicochemical realistic account of what water is, believing that it consists of hydrogen and oxygen and that their combination in water molecules is something that exists in the natural world mind- and theory-independently. What her inquiries focus on is the nature of the religious beliefs and practices surrounding the water taken to be holy. In brief, for the practitioners to believe the water to be holy is not to believe that it does not consist of hydrogen and oxygen, and for the scholar to believe that the practitioners genuinely attach symbolic religious meaning to water is not for him or her to believe that they would believe in the truth of any pseudo-scientific theory (at least not in a realistic sense).

Now, what is the object of such inquiry within religious studies—the object about which we may, in principle, be realists or antirealists (pragmatic or non-pragmatic, practical or nonpractical), more or less analogously to the way in which we can be realists or antirealists about scientific objects? As already remarked, the relevant research object here is not water in the sense in which the scientist deals with water. It is, rather, the symbolic meaning network associated with water in the context of ritualistic practices—and not any water but precisely the specific holy water the practitioners use as a "religious artefact" (employing Kalmykova's term) in their practices—that is the object of the inquiry one engages in within religious studies. The water as conceived of as such an artifact is of course ontologically dependent on our meaning-bestowing interpretive practices of inquiry, but so is (for any Kantian-inspired pragmatic realist, at least) the physicochemical water entangled with our natural-scientific practices.

In principle, the scholar we are imagining could presumably be an antirealist about the chemistry of water when focusing (realistically) on its symbolic meanings, or vice versa. Clearly, the scientist who is a realist about the chemistry of water could easily be an antirealist about those meanings. Whether one maintains a realist or an antirealist interpretation of the properties of water as an object of inquiry depends (among other things) on which discipline the inquiry belongs to and on how one views the realism issue regarding the practices of inquiry in that field.

Tackling these questions further would require that we extend the debate on realism versus antirealism—and the pragmatic and/or practical reinterpretations of this debate—from the general philosophy of science to the philosophy of the humanities, covering disciplines such as religious studies that study meanings and values rather than, say, physical or chemical entities and properties.[11] There is no principled reason why realism could not be developed, with practice-sensitivity and contextualizing awareness, across the board, but this certainly requires further scrutiny. It ought to be investigated how far an individual scholar (or a community) is able to advance antirealism regarding a certain practice-laden theory and its ontological commitments (e.g., the religious studies theory about the believers' views on the properties of water they believe to be holy) while maintaining realism about another, yet partly overlapping, practice-laden theory (say, the scientific chemistry of water). A realist about chemistry might very well be an antirealist about religious studies and the other humanities, claiming that there is nothing in the world "out there" that determines the truth or falsity of our theories in the latter domain and that there are no really existing theoretical entities there that would make our theoretical statements true. (It does not seem unlikely that many scientific realists *are* antirealists in this sense at least about some parts of the humanities, interpreting the practices of inquiry within the humanities very differently from the way they interpret scientific practices.) But could a realist about religious studies be an antirealist about chemistry? That is, could someone coherently claim that the theories of the religious studies scholar do pick out real entities and properties in the world—in this case, the social and cultural world of the religious practitioners' beliefs, rituals, and symbolic meanings—while the theories of the scientific chemist are to be interpreted instrumentalistically or antirealistically (e.g., claiming atoms to be just "useful fictions")? This would presumably be an awkward but not incoherent view to take. At any rate, a more natural and pragmatically viable choice would, I suppose, be to maintain (practical) realism as widely across the interdisciplinary board as possible. When doing so, one just has to recognize how different the practices of inquiry and their theoretical commitments are in different fields—such as chemistry and religious studies. Again, our realism, if pragmatic, needs to be pluralistic and contextualizing.

As noted, Elena Kalmykova's (2024) examination of the profoundly practice-laden character of religious belief is highly relevant here, though I am of course only referring to her work as a case study I am employing for my own purposes. Kalmykova finds the traditional propositional understanding of belief—widely presupposed in the philosophy of religion especially in the Anglo-American analytic tradition—insufficient to account for the complex ways in which religious people's beliefs are intertwined with their practices. She thus argues for an overcoming of the dichotomy between religious beliefs (traditionally investigated in the philosophy of religion) and religious practices (traditionally investigated in empirical religious studies, including anthropology and ethnology), willing to consider religion in its "natural environment" (Kalmykova 2024, 7). She maintains, furthermore, that religious practitioners construct "sacred artefacts"—including both concrete objects such as icons and more abstract ones, such as doctrinal propositions figuring in contexts of worship[12]—and that their relation to the transcendent world they view themselves as encountering in religious activities is predominantly perceptual. That is, Kalmykova somewhat controversially suggests that religious practitioners "perceive" religiously the objects they deal with, and their religious beliefs are realized in embodied actions involving the use of religious artifacts. In particular, embodied religious practices are ways of sustaining the believers' perceptual relations to religious objects that are not "available" due to their transcendent character (Kalmykova 2024, 94). Religious artifacts are needed because the supernatural escapes any ordinary perception—and presupposing this entails more or less taking for granted what the scientific picture of the world says about human perception.[13]

There would be many critical remarks to add on Kalmykova's project—in particular, regarding the normativity and fallibility of our perceptual relation to reality we may (or may not) believe to be transcendent or spiritual—but here I am only using her view as an example illustrating the significance of practical/pragmatic realism in the humanities. While Kalmykova does not explicitly deal with the realism issue in any great detail, she does claim religious believers to postulate certain kinds of cultural objects, that is, sacred artifacts that carry religious and/or spiritual symbolic meanings within the practices the believers engage in. The scholar, on her account, investigates such beliefs and practices (with no essential division between them) and the believers' perceptual activities based on those practices (viz., activities supporting those beliefs), aiming at a true (or truthlike, or at least warrantedly assertible) scientific *cum* scholarly theory of these research objects. The water believed to be holy is, again, a case in point. As emphasized above, the scholar may maintain—as a part of their interpretation of what is going on in the religious practices studied—that religious believers need not believe any nonscientific theories about the chemical structure of water but can follow modern science. However, in addition, the scholar might find out that they postulate cultural and spiritual properties of water while also believing it to be in another sense just natural water. These beliefs are inseparably entangled with the believers' perceptual practices.[14]

Now, does the scholar investigating a group of religious practitioners and their habits of action regarding the water they take to be holy have to have a realistic theory not only of the symbolic meanings (or other cultural objects postulated) but of *perception*, in addition to having a realistic theory of the chemistry of water? Insofar as embodied perception plays a crucial role in Kalmykova's (meta-)theory of religious studies, we presumably must rely on a theory of the ontological status of what goes on in perception and perceptual activities. Insofar as that theory is based on what science tells us about perceptual processes, does the emphasis on perception here contradict the realism about the cultural properties of "holy" water, which, presumably, cannot be "perceived" in the sense in which we perceive ordinary everyday objects, or natural-scientific ones, for that matter—or can one, again, maintain realism across the board?

I am not seeking any definite answers to these questions. What I am suggesting is that Vihalemm's practical realism might turn out to be very helpful here, because it contextualizes any realism we may be able to defend in the practical (including perceptual) activities underlying our theoretical claims and beliefs. Kalmykova's embodied and perceptual theory of religion and religious studies can very well be accounted for in terms of Vihalemm's practical realism, taking seriously the practices of (i) the religious believers studied, (ii) the religious studies scholar studying them, and (iii) the scientist studying the scientific facts that the religious studies scholar must presuppose in their research, including facts about perception and water. Given that any ontological identities are, for pragmatists and practical realists alike, practice-embedded, it is of utmost importance to specify which (possibly overlapping) practices we are operating within when conceptualizing any piece of reality as something specific (e.g., as water or as "holy water").

REFLEXIVITY

Neither Kalmykova nor Vihalemm is prepared to give their positions an explicitly pragmatist interpretation (cf. Kalmykova 2024, ch. 3). This, however, is exactly what I propose to do. A pragmatist account of the realism we are able to defend along practical-realist lines is needed precisely because what I am calling pragmatic realism can (unlike "practical realism") offer a transcendental account of the dependence of ontology on practices in a way "merely" practical or perceptual realism in my view cannot. In other words, as tentatively suggested above, pragmatism is needed here due to the essentially *reflexive* (self-reflective) character of the realism issue when practically conceived. A religious studies scholar focusing on believers' perceptual activities needs to have a theory of perception that is applicable not only to the perceptual activities of the believers but also to his or her own empirical methods of inquiry into the perceptual activities he or she is studying. The scholar cannot just take perception as granted but must subordinate it to scientific and philosophical scrutiny. Similarly, the philosopher of chemistry who takes a realistic attitude to chemical theory also needs, at the meta-level, a background philosophical theory of perception as applied to the chemical laboratory experiments justifying the theories in the field. This reflexivity is duly recognized in practice-based philosophy of science: we are challenged to understand our own scholarly and scientific activities as natural human practices taking place in the world we live in. Indeed, our engaging in such practices is part of our fully natural cognitive life in the natural world. Whatever realism we are able to ascribe to the theories emerging from those practices of inquiry, particularly their ontological postulations (which tell us what the world we take ourselves to be perceiving and theorizing about is like, and what exists) must extend to a realistic account of our practices of inquiry themselves. (Recall: our contextualizing activities themselves only take place in contexts—and this is again something that the pragmatist takes very seriously.)

As already pointed out in the previous section, what troubles me in "mere" practical realism (without full-blown reflexive pragmatism) is the attempt to view scientific and (by extension) scholarly practices from sideways on, so to speak. Pragmatism in the full-fledged sense appreciates our always already finding *ourselves* within those, or at least some, practices. This is precisely why I am defending a "transcendental" pragmatism: we must investigate the ways in which our practices themselves, as seen from within our being (already) committed to them (viz., being practitioners within them), presuppose certain conditions without which they would not be so much as possible for us.

The religious studies example we have briefly examined reminds us about the need for multiple layers of practice-based (practical, pragmatic) realism in the sciences and the humanities. Consider, once more, the water example. We may examine realism about physics and chemistry (or science generally), asking whether the chemical properties of water postulated by our most advanced theory are mind- and theory-independent. We may also examine the religious practitioners' embodied perceptions of water in the context of their rituals. Here the realism issue may concern the psychological reality of their mental states and perceptions, but this is not yet to explore the realism issue as it concerns the water they perceive and use as a religious object or sacred artifact (in Kalmykova's sense, as described above). The realism issue concerning the religious studies scholar's theoretical and empirical account of such objects/artifacts is a further layer in this realism problematic. Here the question concerns the reality of distinctively religious (or more generally symbolic and cultural) meanings and values that the believers, according to the scholar's theory, in their ritual practices attach to the objects they construct and maintain, as well as the ways in which those meanings and values are based on the artifact-employing perceptual processes they engage in. Religious studies—a multidisciplinary inquiry into human beings' ways of living and thinking religiously—may itself be interpreted realistically or antirealistically, like any other human inquiry, and, in this case, the realistic interpretation suggests that the religious practitioners' meanings and values associated with sacred artifacts qua entities postulated in religious studies scholarship are "really there" as elements of the reality investigated. Such meanings and values are not mind-independent, of course, insofar as they are created by human beings due to their participation in religious practices, but they may be completely independent of the researchers' minds and theories.[15]

The kind of religious studies practice of inquiry focusing on the embodied perception inherent in religious practices, as articulated by Kalmykova, must rely not only on a humanistic and social-scientific understanding of practices (as involving meanings, values, and purposes, as something that our humanistic and social-scientific theories and interpretations speak about—and as something that can be interpreted either realistically or antirealistically) but also on a natural-scientific understanding of what embodiment and perception are as material and psychological states or processes. For example, following

Kalmykova (2024), we may say that a religious studies scholar can postulate ("humanistically") sacred objects to which (the scholar claims) religious practitioners attach distinctively religious meanings and can thereby explain the reality (within religious practices) of such objects by referring to the central role played by perception in religious activities and by the fact that sacred objects in a way represent the transcendent objects that remain unavailable in perception. Therefore, I have argued that in religious studies (as understood along these lines) we need a natural-scientific account of perception, including the fact that we *cannot* (naturally) perceive "genuinely" religious entities, "really" transcendent objects beyond the spatiotemporal natural world our perceptual capacities are restricted to.[16] Without such a scientific understanding of perception, we cannot construe the embodied perceptual dimensions of religious practices in the way Kalmykova suggests we ought to do when pursuing religious studies (or trying to understand it philosophically). Therefore, the realism issue may be raised also at the level of the science of perception grounding the perception-centered conception of religious studies, and thus also grounding the specific account of religious artifacts as cultural objects that our inquiries in religious studies may (or, according to Kalmykova, should) yield.

This example enables us to conclude that the various dimensions of realism (or the problem of realism) that may be actualized at different levels and contexts of inquiry are entangled in complex ways. A practical realism along the lines of Vihalemm (2011, 2012) needs to appreciate this radical contextuality of the problem of realism, precisely because it takes seriously the practices that function as contexts for the emergence of the scientific and/or scholarly postulation of any objects of research. The problem, however, is precisely the impossibility of viewing the practice-laden contextuality of realism(s) "from above," from a standpoint beyond all practices of inquiry. For us, there is no such standpoint any more than there is any genuine transcendence in a realistic sense (as distinguished from culturally postulated transcendence we may reach toward through perceptual religious practices employing "religious artifacts").

It is precisely such pluralism and contextualism about the practices of inquiry enabling the real objects of inquiry to be what they are for us that invites a pragmatic (as distinguished from Vihalemm's merely practical) realism, because pragmatism can, as I have suggested, operate here in a manner fully conscious of its Kantian roots. Our pragmatic engagement with realism—our pragmatic realism—must be transcendental because it must be resolutely reflexive. It must recognize that the emergence of the reality we investigate within our scientific and scholarly practices is something that takes place from within the practices to which we cannot adopt a higher vantage point precisely because we are always already within those practices.

Here it is impossible to go into details about this type of pragmatic realism, let alone the history of pragmatist philosophy of science and inquiry (cf., again, Pihlström 2022), but a crucial element of this realism, analogous to Kant's empirical realism, is that it is itself possible only within a transcendental pragmatism (performing the function of Kant's transcendental idealism in a naturalized setting).

The reason I still prefer what I have earlier called pragmatic realism to Vihalemm's practical realism is thus not the willingness to avoid the kind of naturalism (or even materialism) Vihalemm subscribes to. Pragmatism *is* a form of naturalism, too, albeit thoroughly non-reductive. The reason for my preference is the need to develop a truly reflexive form of (practical) realism within pragmatism, and this, I believe, requires a Kantian-inspired transcendental account. Pragmatic reflexivity, pluralism, and contextualism are here entangled, and appreciating their entanglement motivates a transcendental analysis of practices as conditions for the possibility of any scientific and humanistic objects of inquiry.

DEVELOPING PRAGMATIC REALISM IN THE PHILOSOPHY OF THE HUMANITIES

Scientific and scholarly (or any) practices are, indisputably, social and humanly created entities, or complexes of entities and processes. A pragmatic realism of the kind sketched above—as well as the practical realism Vihalemm defends—thus needs to take seriously the reality of practices themselves as the contexts that enable the objects postulated in inquiry to be real, or to be even possible objects of inquiry for the inquiring beings we are. This is one, though of course not the only, reason to examine the realism issue as it arises not only in the philosophy of the sciences but also in the philosophy of the humanities. Our simple example adopted from religious studies has, I suppose, reminded us that the ontology of the practices under scrutiny in a humanistic (or social-scientific) inquiry as well as of the practices engaged in by the inquirers in their respective (overlapping) fields needs serious discussion from the point of view of the realism issue.

The ontology of humanistic research objects (theoretical entities) is also philosophically important in its own right, certainly no less pressing an issue than the ontology of scientific research objects. Instead of electrons, black holes, or the chemical structure of water, we are here talking about objects such as the meanings of historical documents, religious values, or symbolic meanings associated with water believed to be holy (as in our example above), or the meaningful structures of literary works of art—and many different kinds of objects of the same type. As we cannot ascertain the ontological status of these (or any other) entities, except by engaging in normatively guided practices of inquiry within the humanities, our pragmatic realism in the philosophy of the humanities, as much as in the philosophy of science, must be transcendental—or, more precisely, should be formulated, by analogy to Kant, as an empirical realism based on transcendental pragmatism. Let me elaborate on this idea by canvassing the multiple dimensions of the realism issue, now specifically applied to the humanities.

We may chart these dimensions of realism and its alternatives by asking the questions that Ilkka Niiniluoto (1999, ch. 1) finds central in his articulation of *scientific realism* and by bringing them to address the special case of the humanities.[17] *Ontologically*, the realism issue in the humanities concerns the mind-independent existence of the objects of humanistic inquiry (whatever those objects are). The realist about the humanities affirms that at least some of these objects—particularly theoretical entities postulated for theoretical and interpretive reasons within humanistic research—exist and have the properties they do independently of our conceptualizations, inquiries, and theories, analogously to the way in which the scientific realist claims that theoretical entities such as electrons, molecules, or genes exist mind- and theory-independently.[18] *Epistemologically*, the question concerns the knowability (cognizability) of such objects of humanistic research; here the realist claims that we may, at least to some extent, get to know the objects of humanistic inquiry and their properties—though of course it needs to be asked what exactly it means to "know" such things. Such knowledge may, for example, be considered something that consists of interpretive and theoretical understanding of what it means, say, for a certain religious group to ritualistically use water they believe to be holy. The "object" of knowledge or understanding here is this "meaning."

This brings us to the dimension of realism usually labeled *semantic* realism. Here the realism issue focuses on the reference of our theoretical terms and the applicability of the concept of truth to theoretical (and here, specifically, interpretive) statements. According to the realist, the theoretical vocabularies employed by humanistic scholars at least purportedly refer to objects existing in an ontologically realist sense, and the notion of truth can, in principle, be applied to theories in the humanities. That is, according to the realist, our uses of theoretical terms in the humanities either refer or fail to refer to really existing things depending on whether such things exist or fail to exist (independently of individual minds, though not independently of human minds in general), and our statements about those (or any) objects are true or false depending on whether they correspond to the way things are. For example, the religious studies scholar can be (more or less) "right" or "wrong" in their interpretation of the believers' practice-laden beliefs about water.

Furthermore, *methodologically*, we may ask whether it is possible to develop research methods that (tend to) lead us to the truth or that yield an inquiry progressing toward the truth (in a quasi-Peircean sense) in the humanities;[19] again, the realist affirms this, at least in some sense. Finally, there is the *axiological* question about the values, aims, and goals of research: Does humanistic inquiry aim at truth, as the realist claims, or does it pursue something else, such as practical problem-solving or individual or cultural welfare and self-understanding, or merely social or political empowerment?

A full-scale realism about the humanities would, when faced with these questions, state roughly the following. There is a real social and cultural world "out there" independently of individual humanistic scholars' (and research groups') use of concepts, language, and theories. It is, admittedly, a world created by human beings and their activities, containing objects such as meanings, norms, values, actions, and institutions, but the existence and properties of such entities are largely independent of any particular inquirers' or groups' of inquirers' views, beliefs, opinions, and theories. By means of practice-embedded theoretical research and its rational methods of inference, explanation, and interpretation, we may, at least to some extent, get to know and understand this humanly created cultural world more deeply. We may refer to its objects, events, and properties by using theoretical language, and our statements about it are, at least in most cases, true or false depending on the way the world—conceived as consisting of those objects, events, and properties—is. Humanistic inquiry pursues truth about this cultural reality and seeks to develop methods that may bring scholars closer to the truth. All of this takes place within normative practices of inquiry, and our theoretical understanding of human culture depends on such practices. The objects of such understanding, or even their very possibility, also ontologically depend on those practices of inquiry—and as we saw, Vihalemm, indeed, rightly argued that Niiniluoto's realism pays insufficient attention to this practical basis of realism.

Pragmatists may take various stands between realism and antirealism. Arguably, no sane pragmatist will deny realism altogether—either in the sciences or in the humanities. Instead, a reasonable form of pragmatic realism needs to be distinguished *both* from metaphysical realism committed to a "ready-made" world existing with its own pre-categorized ontological structure *and* from radical antirealisms, such as thoroughgoing relativism and/or constructivism that deny the theory-independent reality of any objects of study (in their most radical form even in the natural sciences). Again, in steering this middle course between the extremes of strong realism and antirealism, I believe I am fully in agreement with Vihalemm's practical realism. Pragmatists have defended plausible and sophisticated accounts of pragmatic realism (or realism integrating key insights from constructivism and/or transcendental idealism) in general epistemology and philosophy of science; it is a major further task for pragmatist philosophers of the humanities to investigate how far these conceptions may be applied to the philosophy of the humanities.[20]

In a pragmatist philosophy of the humanities (which cannot be developed at any length here, of course), the ontological question concerning the reality of the objects of humanistic inquiry needs to be examined in close relation to the realism issue in its full diversity—especially in dialogue with the Kantian idea (in my view, inherited by pragmatism) of the transcendental constitution of the objects of inquiry. While the natural sciences study (according to scientific realism) objects, processes, and laws existing and obtaining in the natural world independently of the human mind and of our scientific concepts and theories, and while the social sciences seek to explain and understand humanly created social reality, such as social structures and institutions, it may be suggested that the humanities primarily examine humanly created *meaningful* objects (e.g., texts and historical documents) as well as their historically contextual meanings and representational relations to whatever they may be taken to be "about." However, scholarship in the humanities also in a sense focuses on other portions of reality beyond such texts, namely, things and events that those documents themselves may refer to, such as historical events that "really" took place. Thus, for example, a war historian may examine archival documents and by using them ask (and answer) questions about the historical events to which those documents may be taken to stand in some kind of representational relation. Accordingly, the "object" of such historical research is not simply the document, nor simply the event the document may be thought to (accurately or non-accurately) represent, but such representational relations themselves, as well as, critically, their purported accuracy.[21] The question concerning what the humanities are actually "about" is therefore far more complex than it might prima facie seem. Humanistic inquiry may be "about" historical and meaningful objects and events, but it may also be "about" (the meanings of) documents that purportedly themselves refer to such objects and events, and about those referential links. Similarly, literary criticism may focus on the way a historical novel refers to both historical reality and fictional characters, which makes the ontological status of the objects of interpretation rather complex.[22] And as we saw in our example in the previous section, religious studies may investigate the relationship religious believers' perceived sacred objects have to natural entities (e.g., water), and the ways in which these relationships are relevant to the symbolic meanings the religiously perceived objects may be taken to possess.

The question about the objects of humanistic inquiry, as a manifestation of the question about realism and truth, thus concerns the sense(s) in which those objects, including literary meanings, events and actions in our human past, or religious doctrines manifested in a certain institutionalized practice, are "real" as objects of study—and thus even potentially available for true or false scholarly or theoretical representation—even though to be "real" for pragmatists always means to be embedded in practices of conceptualization and inquiry. As we have seen, it may also be asked whether there are "theoretical entities" in the humanities comparable to the theoretical entities postulated in natural-scientific theories, that is, unobservable entities whose existence explains observable phenomena that would otherwise be very difficult or impossible to explain.[23] My general proposal is that the pragmatist about the humanities should take a basically realistic attitude to the ontological postulations of humanistic theorization from within the practices of that theorization, in the same spirit of pragmatic realism in which s/he would approach the realism issue in general philosophy of science—without, however, claiming such theoretical entities to exist in any "ready-made" world that would be "there" independently of the scholarly practices enabling those theoretical postulations. Both the postulation of theoretical entities and the conception of truth involved in a realistic account of such practice-dependent theoretical discourses as "truth-apt" must ultimately be subordinated to pragmatism.[24]

CONCLUDING REMARKS

Having now examined the issue of (pragmatic) realism in the philosophy of the humanities, we should finally briefly return to Rein Vihalemm's notion of "practical realism," which he invited us to compare to pragmatic realism. Is there, after our analysis, any significant difference between the pragmatic and the practical in this regard?

The philosophy of the humanities can be as profoundly practice based as the philosophy of science, and the realism in the field may be as fully practical realism as Vihalemm's favorite form of scientific realism is. As explained, the key reason I have preferred (my own version of) pragmatic realism to Vihalemm's practical realism is the need to emphasize the reflexivity of our practices of inquiry: it is from the point of view of our practices of inquiry that the issue of realism concerning not only the objects they are about but also the constitutive features of those practices themselves arises. I have briefly described the significance of this reflexivity by drawing attention to the ways in which scholars—such as religious studies scholars in our simple example—may at the same time be involved in a plurality of partly overlapping practices and also have to consider the (realist or non-realist) ontology of their own practice. Yet, I have not shown pragmatic realism to be superior to practical realism in any clear sense. On the contrary, I think of these two brands of practice-involving realism as belonging to one and the same research program in the philosophy of science and the humanities. They should be developed together in this spirit.

Finally, in this chapter, I have *not* argued that the pragmatist (or the practically oriented philosopher of science and the humanities) *ought to* focus on the realism issue in the first place. There are certainly pragmatists—especially those following Rorty—who would like to set the entire realism discussion aside, and also many pragmatists critical of Rorty's radical views have argued that the classical pragmatists' (especially Dewey's) views cannot be reduced to any standard opposition between realism and antirealism.[25] I have not engaged with these issues here, because I have done so at considerable length elsewhere—and also because this is not a matter I ever disagreed about with Rein Vihalemm, whose legacy I hope to have honored with my modest contribution. Despite our minor disagreements about practical versus pragmatic realism, as also manifested in this chapter, we always agreed about the fundamental importance of the problem of realism.

NOTES

1. Thanks are due to Ave Mets for the kind invitation to contribute this chapter to this collection, to two anonymous reviewers for useful critical comments (which I have only partially been able to take into account), to Elena Kalmykova for the permission to cite her still unpublished book manuscript on religious practices, and, of course, to the late Rein Vihalemm for a number of memorable conversations on realism over the years.

2. I will elaborate on this realism about the humanities (especially regarding the ontological status of the research objects of humanistic scholarship) below. I was fortunate to have an opportunity to discuss realism, pragmatism, and practical realism with Rein Vihalemm on a number of occasions from the very early 2000s to his untimely death in 2015, but I very much regret that I never properly took up the realism issue concerning the philosophy of the humanities in those discussions.

3. See, for example, Mäki (2005, 2007).

4. See, for example, Pihlström (2021, 2022, 2023b).

5. See also Chang (2022).

6. Here I cannot deal with this complex Kantian issue in any detail, but let me note in passing that the pragmatic realist whose realism is grounded in a transcendental pragmatism (a pragmatist analogy of Kant's transcendental idealism) need not give up realistic or even correspondence-theoretical truth; instead, as William James ([1907] 1975) himself suggested, pragmatism formulates its conception of truth as a specification of what the "agreement" between our beliefs and reality figuring in the correspondence theory actually amounts to. (See further Pihlström 2020, 2021. Also note that I am not citing the pragmatist classics in any detail in this chapter; I have done so extensively in many other publications.)

7. I am here elaborating on an issue briefly raised in my earlier engagement with Vihalemm's views (see Pihlström 2014).

8. I am speaking of "theories" for the sake of brevity here, of course acknowledging (as a pragmatist) their thoroughgoing practice-embeddedness: no theory is "as such" interpretable realistically or antirealistically, but we are always speaking about the complex ways in which the ontological commitments of a theory are themselves irreducibly practice-embedded and practice-laden. Here, again, I am fully with Vihalemm. Accordingly, we should speak about the realistic versus antirealistic interpretations of such complex commitments that involve both theory and practice, but we can presumably stick to the standard way of discussing the interpretations of scientific theories even when acknowledging their ineliminable grounding in scientific practices—and similarly for whatever theories and practices may be relevant in the humanities.

9. When speaking of the "subjective" (practice-based) ontological grounding of whatever "objective" facts of reality there are, the pragmatist obviously does *not* subscribe to any traditional subject–object dualism. Rather, our human practices themselves—scientific and nonscientific—function as the grounding of both the subjective and objective aspects of reality. The reference to "subjective" is thus to be understood as shorthand for something that is not taken to be simply objectively given independently of our contexts and perspectives but is constituted through them. Among the many useful pragmatist analyses of the issue of realism and the overcoming of the subject–object dichotomy, see, for example, Hildebrand (2003).

10. Note that for the (e.g., Peircean or Deweyan) pragmatist, even the purest theoretical investigation of the structure of water is never just an inquiry into what water "is" as something absolutely independent of us but always also an inquiry into what we—within and through our practice—"do" with water. (Here I am grateful to the anonymous reviewer's helpful suggestion.)

11. Any pragmatist or pragmatically realist study of "meanings" must, of course, begin with the idea of meaning being grounded in "use" (loosely employing Wittgenstein's famous words), and the classical pragmatists' different versions of the Pragmatic Maxim may be regarded as variations of this theme. See, for example, the 1878 essay "How to Make Our Ideas Clear" in EP 1 and James (1907) 1975, lecture II. I deal extensively with these topics in many of my writings listed in the bibliography, most recently in Pihlström (2023a). I do acknowledge (again in dialogue with one of the reviewers) that the practice-based character of meanings may make the comparison with theoretical entities in science problematic. On the other hand, neither the historical meanings of a document nor the subatomic particles that physical objects consist of are immediately observable; I do take this to be a relevant, though not complete, analogy.

12. See Kalmykova (2024, ch. 2).

13. Clearly, the religious studies scholar—in order to maintain scientific credibility—cannot claim the religious practitioners to be able to literally perceive anything transcendent when engaging in their religious practices and using their sacred artifacts within those practices. Nevertheless, one of Kalmykova's (2024) key points is that perception, when accounted for in a habitual and embodied sense, is relevant to

their activities. I cannot evaluate the credibility of Kalmykova's position as a contribution to religious studies; I am merely citing it as a philosophically interesting example of the kind of issues that pragmatists and pragmatic realists may face when extending their realism from general philosophy of science to the philosophy of the humanities.

14. In Wittgensteinian terms, we could say that they see the water "as" containing mysterious transcendent powers even while not disagreeing with the chemical theory of water. Compare Kuhn's ([1962] 1970) scientists working within different paradigms: they may see something entirely different when looking at the same scientific data, and there is no easy or straightforward sense in which they can even be said to disagree with each other because agreement and disagreement are only possible within a paradigm.

15. A further issue of realism concerns the interpretation of religious discourses and practices themselves, as distinguished from the discourses and practices of religious studies. One might interpret religion itself antirealistically while interpreting religious studies realistically—or, in principle, though hardly ever in practice, vice versa. Another complication is the distinction between theology and religious studies, which also takes different shapes in different cultural contexts, but I will neglect that distinction here. (For a more comprehensive discussion of these issues, see Pihlström 2020.)

16. For this reason, many religious studies scholars and philosophers, and scientists for that matter, maintain that there are no such objects at all, or at least that there is no reason to believe that there are any. This question, however, falls outside the scope of this chapter, which is not directly concerned with the philosophy of religion.

17. See also Livingston (1988).

18. Again, recall that, clearly, the objects of humanistic inquiry do *not* exist independently of human minds, precisely because they are (typically) humanly created cultural entities of some kind. Characteristically, these objects may not exist theory-independently either, because it may be meaningful to discuss their existence and their properties only within a certain theoretical framework (though in a sense, albeit perhaps not exactly in the same sense, this can also be claimed to be the case with natural-scientific theoretical entities). However, whatever theoretical entities are postulated in the humanities, they may still be, realistically, independent of the researchers' individual or collective opinions, beliefs, and wishes: they are not simply subject to any of our contingent subjective ways of thinking or talking about them. Moreover, realism does not depend on any particular objects existing or having the characteristics they are taken to have; both scientific and humanistic realism are fully compatible with the progress of inquiry continuously correcting our picture of *which* theoretical entities exist and what properties they have.

19. Famously, Charles Peirce, in "How to Make Our Ideas Clear" (1878, in EP 1) and elsewhere, characterized truth as the "final opinion" toward which inquiry *would* converge if it were continued indefinitely long (cf., e.g., Misak 2013).

20. See, again, Pihlström (2022). I might note that my pragmatic realism is most deeply indebted, in addition to the classical pragmatists, to Putnam's (e.g., 1990, 2016) struggles with realism over the decades. Putnam, however, never seems to

have explicitly (apart from scattered remarks here and there) applied his "internal realism" (his view in the 1980s) or his later versions of pragmatic realism to the humanities—any more than Vihalemm ever applied practical realism to the humanities. (It is not my task in this chapter to extensively refer to pragmatist literature or to trace the various sources of the kind of pragmatic realism I am merely sketching here.)

21. The issue of representation is of course complex here, and invoking the concept at all presupposes that we maintain a critical distance to Richard Rorty's (e.g., 1998) radically antirepresentationalist pragmatism. Of course, it is misleadingly simple to say that historical documents "represent" historical facts, but there needs to be some relevant relation—within a normatively structured human reality of meanings—between, say, an archival document and some real historical event in order for the historian's research questions to make sense. Regarding pragmatist views on representation, Peirce's semiotics would obviously also be highly relevant but must unfortunately be neglected in this investigation.

22. A historical novel typically sets fictional characters and events in "real" (or "truthlike") historical circumstances. However, in a somewhat more complex case, fictional characters can be set in a counterfactual historical situation that retains some elements (e.g., people) from "real" history, as in Philip Roth's *The Plot Against America* (which adds the further complication of integrating fictional events with a semiautobiographical background).

23. For a lucid discussion of the reasons to postulate theoretical entities, see again Niiniluoto's (1999) defense of scientific realism.

24. See, again, Pihlström (2021, 2022).

25. See Rorty (1998) and Hildebrand (2003).

REFERENCES

Chang, Hasok. 2022. *Realism for Realistic People: A New Pragmatist Philosophy of Science*. Cambridge: Cambridge University Press. https://doi.org/10.1017/9781108635738.

Dewey, John. [1929] 1960. *The Quest for Certainty: A Study on the Relation Between Knowledge and Action*. Boston, MA: G. P. Putnam's Sons.

Hildebrand, David L. 2003. *Beyond Realism and Antirealism: Dewey and the Neopragmatists*. Nashville, TN: Vanderbilt University Press. https://doi.org/10.2307/j.ctv16b78d6.

James, William. [1907] 1975. *Pragmatism: A New Name for Some Old Ways of Thinking*, edited by Frederick H. Burkhardt, Fredson Bowers, and Ignas K. Skrupskelis. Cambridge, MA and London: Harvard University Press. https://doi.org/10.1037/10851-000.

Kalmykova, Elena. 2024, forthcoming. *Bridging Beliefs and Practices*.

Kuhn, Thomas S. [1962] 1970. *The Structure of Scientific Revolutions*. 2nd edition. Chicago and London: The University of Chicago Press.

Livingston, Paisley. 1988. *Literary Knowledge: Humanistic Inquiry and the Philosophy of Science*. Ithaca, NY and London: Cornell University Press. https://doi.org/10.7591/9781501746024.

Mäki, Uskali. 2005. "Reglobalising Realism by Going Local, or (How) Should Our Formulations of Scientific Realism Be Informed About the Sciences." *Erkenntnis* 63: 231–51. https://doi.org/10.1007/s10670-005-3227-6.

Mäki, Uskali. 2007. "Putnam's Realisms: A View from the Social Sciences." In *Approaching Truth: Essays in Honour of Ilkka Niiniluoto*, edited by Sami Pihlström, Panu Raatikainen, and Matti Sintonen, 295–306. London: College Publications.

Misak, Cheryl. 2013. *The American Pragmatists*. Oxford: Oxford University Press.

Niiniluoto, Ilkka. 1999. *Critical Scientific Realism*. Oxford: Oxford University Press.

Peirce, Charles Sanders (EP). 1992–98. *The Essential Peirce. Selected Philosophical Writings*. 2 volumes. Vol. 1 edited by Nathan Houser and Christian J. W. Kloesel; Vol. 2 edited by The Peirce Edition Project. Bloomington, IN: Indiana University Press.

Pihlström, Sami. 2014. "Pragmatic Realism." In *Realism, Science, and Pragmatism*, edited by Kenneth R. Westphal, 251–82. London and New York: Routledge.

Pihlström, Sami. 2020. *Pragmatic Realism, Religious Truth, and Antitheodicy: On Viewing the World by Acknowledging the Other*. Helsinki: Helsinki University Press. https://doi.org/10.33134/HUP-2.

Pihlström, Sami. 2021. *Pragmatist Truth in the Post-Truth Age: Sincerity, Normativity, and Humanism*. Cambridge: Cambridge University Press. https://doi.org/10.1017/9781009047142.

Pihlström, Sami. 2022. *Toward a Pragmatist Philosophy of the Humanities*. Albany: SUNY Press.

Pihlström, Sami. 2023a. *Humanism, Antitheodicism, and the Critique of Meaning in Pragmatist Philosophy of Religion*. Lanham, MD: Lexington.

Pihlström, Sami. 2023b. *Realism, Value, and Transcendental Arguments Between Neopragmatism and Analytic Philosophy*. Cham: Springer. https://doi.org/10.1007/978-3-031-28042-9.

Putnam, Hilary. 1981. *Reason, Truth and History*. Cambridge: Cambridge University Press. https://doi.org/10.1017/CBO9780511625398.

Putnam, Hilary. 1990. *Realism with a Human Face*, edited by James Conant. Cambridge, MA and London: Harvard University Press.

Putnam, Hilary. 2016. *Naturalism, Realism, and Normativity*, edited by Mario de Caro. Cambridge, MA and London: Harvard University Press. https://doi.org/10.4159/9780674969117.

Rorty, Richard. 1998. *Truth and Progress*. Cambridge: Cambridge University Press.

Rouse, Joseph. 2002. *How Scientific Practices Matter: Reclaiming Philosophical Naturalism*. Chicago, IL and London: The University of Chicago Press.

Vihalemm, Rein. 2011. "Towards a Practical Realist Philosophy of Science." *Baltic Journal of European Studies* 1 (1(9)): 46–60. Accessed October 2023, http://www.ies.ee/iesp/No9/iesp_no9.pdf.

Vihalemm, Rein. 2012. "Practical Realism: Against Standard Scientific Realism and Anti-Realism." *Studia Philosophica Estonica* 5 (2): 7–22. https://doi.org/10.12697 /spe.2012.5.2.02.

Westphal, Kenneth R., ed. 2014. *Realism, Science, and Pragmatism.* London and New York: Routledge. https://doi.org/10.4324/9781315779515.

Chapter 8

The Philosophical Neglect of Chemistry Revisited

Klaus Ruthenberg

In the early 1990s, I entered the then very small international group of philosophers of chemistry with the first versions of my interpretation of the philosophical neglect of chemistry.[1] The first humble attempt was the short presentation, "Is There a Philosophy of Chemistry?" at the ninth International Congress of Logic, Methodology and Philosophy of Science in Uppsala on August 10, 1991. It was the first time I met Rein Vihalemm, who criticized my merely negative answer to the title question of that talk.[2] Since then, we met on several occasions. The present contribution does not directly refer to Vihalemm's work. However, it might be considered as a supplement to the general view he called practical realism. Particularly, his conviction that the real world can only be made accessible by practice appeals to me because the latter fits chemistry in many ways (cf. Vihalemm 2012).[3]

The present contribution takes a closer look at the identity of chemistry and the question of its still customary philosophical neglect. Among the variety of reasons for this neglect, the ignorance of the substance notion and the overemphasis of submicroscopic entities are the most prominent.

DEFINITIONS AND PLURALISMS

There are good reasons to consider chemistry as a pluralistic endeavor.[4] The most important of these reasons certainly is the fact that the scientific description of substances throughout suffers from epistemological and ontological underdetermination.[5] Even substantial entities with a "sturdy envelope" like water are still subjects of ongoing research.

Many contemporary chemists would perhaps claim that most chemical phenomena can be explained based on atomic physics, and quantum mechanics,

in particular. This is at least the impression one can get from public representations of chemical facts and activities in textbooks and by authorities and
professional societies. Moreover, several philosophers of mind seem to think
that the actual chemical knowledge allows for the reductive explanation of
mental states and perhaps even emotions.

The self-description of chemistry has gone through a few main stages. The
chemist and historian of chemistry Georg Lockemann provided the coherent
observation: "It was only very late that the realization gained acceptance that
the actual task of chemistry is to research the characteristics of the various
substances and their reciprocal transformations. This occurred in the 17th
century" (Lockemann 1950, 5).

Although the concept of substance itself requires clarification, this characterization initially sounds quite plausible. One might argue for an earlier
beginning of (practical) chemistry, but it is certainly correct to point out that
modern chemistry as a scientific discipline started in the decades (or century)
before the long nineteenth century. The definition of chemistry as the science of substance changes is the earliest type of self-description which can
still claim wide consensus, albeit not always explicitly. I concur here with
Elisabeth Ströker, who says something very important about general patterns
of thinking in chemistry: "These patterns of thinking . . . are never plainly
and simply something past: They are also reflected in the current methods of
science, even stratified in their contemporary concepts" (Ströker 1968, 747).
I will call the depiction that chemistry deals with the qualities of stuff "type
1." The historical follower of the stuff-centered definition of type 1 is the
compositional definition, which was an invention of the nineteenth century.
"Compositional" means that with the introduction of the modern concept
of element by Lavoisier, the internal structure of material samples became
the prevailing interest of chemists. As the modern element concept gains
acceptance, self-descriptions begin to appear which emphasize the chemical elements—that is, substances that cannot be further divided into other
simple substances. A modern example of this version which I will call "type
2" comes from an introductory university textbook of inorganic chemistry:
"Chemistry is a natural science. It concerns itself with the study of the human
environment accessible either directly via the senses or indirectly via suitable
instruments. . . . Chemistry is thus the science that concerns itself with the
possible combinations of the 104 known elements" (Schmidt 1967, 1).

At best, as in the cited example, type 2 definitions take for granted that any
chemical investigation starts with some sample from our surroundings. In
general, and in sharp contrast to the applications in the manifest world, however, substances (taken as research subjects) have vanished more and more
from scientific attention. Increasingly, chemists emphasize electrons, atoms,
and molecules, and, intriguingly, in contemporary philosophy of chemistry,

the reference discipline is—wrongly—described as "molecular science."[6] This becomes even more obvious when we recognize what I call the type 3 definition of chemistry, which historically belongs to the twentieth century. According to that type, the world is just made of elementary particles, and chemistry is supposed to uncover the kind, number, and relations of these particles. One example of type 3 from a well-known German chemistry lexicon is the following: "Chemistry is the science that concerns itself with causes and effects of the release, acceptance, or distribution of electrons between atoms or molecules and with the relationships between the energy levels of such electrons within the atoms or molecules" (Römpp 1975).

Similar statements can easily be found elsewhere. Modern analytical chemistry, for example, is often presented without any reference to substance. Representative of the tenor of a large part of the scientific community (being selected and awarded by a jury of specialists), the winner of a competition, a merited professor of analytical chemistry, maintains:

> Analytical Chemistry is the cognitive sense of every natural science discipline grasping for an unbiased chemical reality of the microcosm, whether it is an isolated sample of matter in our hands or a star far away in the universe. It is the type and number of atoms, together with their arrangement in three-dimensional space, which determine the properties of matter. (Cammann 1992, 812)

Such a statement moves the intrinsic, supposedly unbiased "chemical reality" into the microcosm of the unobservables and turns the existence of atoms and their countability and spatial arrangement into what amounts to a metaphysical prerequisite. The empirically necessary work is subordinated to this "chemical reality" and, at best, appears in such narratives as "sample preparation." In other words, the qualitative and phenomenological analysis is disregarded or simply assumed to be somehow self-evident.

This brief discussion of the characterization of chemistry as a scientific enterprise sets the frame for the main topic the present chapter is devoted to.

CHEMISTRY, A PHILOSOPHICAL STEPCHILD?

In order to get an impression of the current status of chemistry in philosophy, one can have a look at the relevant philosophy of science journals such as *Philosophy of Science*, *British Journal for the Philosophy of Science*, *Synthese*, and *Zeitschrift für allgemeine Wissenschaftstheorie*, as well as dictionaries.[7] Even after what is now about three decades of flourishing professional academic philosophy of chemistry, the result of such a search for articles pertaining to chemistry is sobering. Apart from articles in the two specialized journals (*Hyle* and *Foundations of Chemistry*)[8] and a larger series of omnibus volumes,[9] the reader gets the impression that the traces of philosophy of chemistry within the general philosophy of science have to date been relatively few and far between. This feels disproportionate particularly in light of the fact that the scientific output and societal relevance of chemical knowledge far outweigh those of all other areas of science. It would seem that "both" sides have tended to keep to themselves, possibly due to the high degree of technical specialization (see below). Even in dictionaries, which are less dependent on current developments and are thus in a certain sense historically more stable, chemistry as a reference science continues to fare poorly. The renowned *Historisches Wörterbuch der Philosophie*, for example, devotes nine columns of text to physics and one to biology, whereas there is no such entry as "chemistry." Therefore, the following question continues to be of fundamental interest for the philosopher of chemistry: What exactly is it that has, at least to date, made the study of substances appear so uninteresting or even irrelevant in the eyes of philosophical communities? Hopefully, addressing this question will also help to clear up the problem of what actually distinguishes chemistry, what are its crucial characteristic traits. The literature reveals several different approaches to answering the question posed here. The most important of these approaches are the immaturity thesis, the complexity thesis, the reducibility thesis, the dematerialization thesis, and the relevance thesis. In the following, I will discuss these theses in that order. The authors mentioned here call these theses. They are not necessarily proponents or even advocates of the ones I have summarized here for the purpose of discussion.

IMMATURITY

The immaturity thesis assumes that a natural science becomes interesting for philosophy only in an advanced stage. Peter Janich (1942–2016) distinguishes between three graded suitabilities for disciplines to develop into sciences and states:

> The discoveries of *theoretical suitability, empirical suitability,* and *philosophical suitability* [*Theoriefähigkeit, Empiriefähigkeit, Philosophiefähigkeit*] of the natural sciences benefit physics in each case (and later biology as well) but disadvantage chemistry. Thus in whatever causal relationship to one another they may stand, chemistry initially lacks both academic reputation and academic institution. (Janich 1992, 65–66; Janich's emphasis)

From this perspective, chemistry as a modern science appears to be still too young for philosophical reflection (Janich 1980, 1992; Liegener and Del Re

1987). A comparison with other scientific fields is interesting. If we take a look at the biological and psychological disciplines, it may be assumed that the respective underlying sciences are not further developed or more mature than chemistry. And yet, certain aspects that can be associated with them have been subjected to very intensive philosophical examination. Examples of these are the (historical) life-force discussion, the problem of evolution, the question of life in the strict sense, and the mind-body problem. One may also ask why physics questions, if we may call them such, were of interest to philosophers like René Descartes and Immanuel Kant long before the emergence of modern theories such as the theory of relativity and quantum theory. A more substance-related doctrine based on mutability and preparation would not have been excluded next to the more abstract mechanistic notions. Therefore, scientific maturity appears to me to play less of a role in the "philosophical suitability" of a scientific undertaking.

COMPLEXITY

The complexity thesis maintains that chemistry's complexity and methodological diversity, together with its confusing history, render philosophical treatment difficult or even impossible. In other words, the experts in philosophy of science and chemistry have not yet grasped the specifically philosophical problems of chemistry (Ströker 1967; Plath 1990). If complexity and diversity were to prevent a fundamental philosophical reflection in the chemical realm, then that would pose the question of why it is that chemists themselves are well versed in their field but are hardly inclined to reflect on its fundamentals. Specialists in the sciences of substances are generally regarded as decidedly anti-philosophical. Regarding that point, we have to exclude earlier chemists, particularly the alchemists, like Robert Boyle, Andreas Libavius, Isaac Newton, and Daniel Sennert. For those scholars, epistemological and ontological purposes were generally intertwined, and the researcher himself was always thought to be personally involved in any activity in the workshop or laboratory. Modern science cuts this connection, and one interesting result as to the present discussion is that even those professional philosophers who had a first training as chemists did only rarely or occasionally, if at all, think about their first disciplinary realm. Hans Cornelius, Werner Leinfellner, Grover Maxwell, Hans Sachsse, and David Theobald come to my mind here, and only the last two have written significantly about chemistry. Fortunately, the list of scholars formally trained in both disciplines has grown during the past four or five decades. There is Robin Hendry, Jean-Pierre Llored, Paul Needham, Nikos Psarros, Joachim Schummer, Jaap van Brakel, and Rein Vihalemm, among others. Chemical complexity did not prevent quite a number of successful chemists without formal philosophical training, such as Joseph Earley, Roald Hoffmann, Alwin Mittasch, Wilhelm Ostwald, Friedrich Paneth, and Hans Primas, from thinking also about the foundations of their own discipline. Nevertheless, chemistry has known only one single representative of its own guild who developed, in full scope, a philosophy of chemistry—Wilhelm Ostwald.[10]

At first glance, it is hardly understandable why the complexity of a problem area should discourage an inclined philosopher from familiarizing herself with it. Other contexts are often not readily amenable to comprehension and yet attract attention, possibly for that very reason (how many hundred philosophical studies are there on quantum mechanics?). Is the philosophical status of chemistry thus so pitiful because not enough time has passed in order to find attractive theories, or are there, for reasons yet unknown, none to find? Those who primarily seek theories will have to conclude that chemistry actually exhibits a pluralistic diversity and that it therefore requires a different approach than, say, physics. To overstate the case, one can say that chemists do not search for theories in chemistry, rather they hunt for synthetic success and success in a wide variety of applications. To this extent, the complexity theory is indeed significant because it reveals that chemistry is different from mathematical physics. Without a change in perspective—away from static theoretical corpuscular thinking and toward dynamic substance-oriented thinking—one will not be able to do justice to the former.

REDUCIBILITY

The prejudice or, respectively, the conception that chemistry can at least theoretically be reduced to physics (theory reductionism) is widespread in broad areas of the natural sciences and, unfortunately, in the current philosophies of science as well. In the form of a reducibility thesis, the only aspect of chemistry that is of philosophical interest is that which is already being examined within the philosophy of physics (Primas 1985; Liegener and Del Re 1987; van Brakel 2014). The at times lengthy discussion as to whether chemistry is theoretically reducible to physics, a discussion that has accompanied the recent flourishing of the philosophy of chemistry from the outset, is one I cannot describe here. This discussion has, unfortunately, focused more on what chemistry is not and has itself placed disproportionate emphasis on the conventional physics-centric view of science. Thus, too many publications ostensibly in the philosophy of chemistry have, by their own choice, placed themselves in the position of having to justify their existence. In chemistry, it is not a matter of achieving a high degree of mathematization or physicalization, although both need not be detrimental; rather, it is a matter of controlling the substance environment, both the natural one and the man-made one. As a result, I prefer the conventional study-of-substance (type 1) definition, for example, as formulated by Wilhelm Ostwald:

> As the quantity and external shape of a body belong to its arbitrary character-
> istics that can be changed at will, one observes in chemistry the bodies without
> regard to quantity and form. Bodies that are observed in such a sense with
> respect to their specific characteristics are known as substances. It is thus the
> substances that form the object of chemistry. Those substances that have match-
> ing specific characteristics are said to be identical. (Ostwald 1907, 6–7)

Although not all epistemic objects in chemistry are of a substantial nature, the substance concept is the central reference point, and substances can hardly be defined by strictly physical means. The latter might help enormously, but chemistry is no φ-science, as Rein Vihalemm put it some decades ago.

Supported by the main argument that even broad areas of physics itself are not reducible to any fundamental theory, I will limit myself, for the time being, to accepting the thesis that whereas chemistry should be theoretically compat-ible with physics, it is not reducible to the latter. If the reductionist approach were a sufficient answer to the question as to the reason for the deficient philosophical status of chemistry, then the consequence would be that it would be classified similarly to such fields as mechanics, acoustics, thermodynam-ics, the theory of electricity, and maybe fluid dynamics, and meteorology—as philosophically irrelevant. There are certain areas within modern physics that imply epistemological and ideological questions for many philosophers, for example, the theory of matter (particle theory), theory of relativity, and quantum theory and mechanics. In fact, quantum mechanics provides the con-ceptual framework for the modern theory of chemical bonds (cf. Ruthenberg 2020b). However, there is no discussion of problems of quantum chemistry in the current philosophy of science. Reductionism is often encountered in one form or another, especially in physics. However, hardly anyone explicitly sup-ports the opinion that its (assumed) reducibility makes chemistry philosophi-cally irrelevant. A causal relationship between reducibility and philosophical unsuitability does not appear to exist, as other philosophically interesting aspects are also subject to the assumption of reducibility with respect to other disciplines (such as biology and psychology). This thesis thus lacks the possibility of explaining the motivation for biological-philosophical and psychological-philosophical connections. Furthermore, it would not explain why chemistry outside of an assumed (current) reduction phase—disregarding for the moment the existing exceptions—has been a philosophical stepchild. The fact that experts from the sciences of substances occasionally apply reduc-tionist arguments in their own explanatory conceptions and then immediately switch back to talking about macroscopic substance characteristics certainly does not help the situation. In any case, the reducibility thesis by itself does not sufficiently explain the philosophical undervaluation of chemistry.

DEMATERIALIZATION

Joachim Schummer advocates what I, again simplifying, call the *dematerialization thesis*, which he draws from his meticulous study of the history of philosophy, especially the history of natural philosophy and the philosophy of science. His interpretation is that during the course of the history of philosophy, which had its origins in the philosophical examination of substances, the material character of the world was later suppressed and the formal aspect (or φ-scientific aspect) overemphasized. The science of the substances thus does not come to bear as a reference science in philosophy because its object is practically unknown there (Schummer 1996). Schummer sees the beginning of dematerialization with Parmenides. In a certain sense, Parmenides turns Heraclitus' experiential ideas conceptually upside down. Elisabeth Ströker, whose *Denkwege* continues to represent a standard work in early substance philosophy, expresses this aptly: "Whatever is not being, is not. The manifest diversity and manifoldness of existence thus sinks before Parmenides' thinking into appearance" (Ströker 1967, 22). Schummer describes the situation with a startlingly simple example, a sphere of water: the form philosopher will give (ontological) priority to the spherical shape, the substance philosopher to the liquid material, water. While for the former the idea of the spherical shape is there first, the latter sees the substance with its specific characteristics (Schummer 1996, 143–44). Schummer advocates the view that the form approach has prevailed in the history of philosophy and that the substance efforts have been pushed "into an esoteric corner" to this day (Schummer 1996, 144). In his view, it is particularly the three old views about ignorance described by the pre-Socratic philosopher Gorgias that play a role in answering the question of philosophical disinterest in chemistry: the ontological view (there are no substances), the epistemological view (we cannot detect them), and the view of the philosophy of language (if we could detect them, we could not speak about them). Added to this is the fourth attitude, prevalent in the current philosophy of science: "If there were a science that claims to speak about substances, then it would only be speaking about forms" (Schummer 1996, 152).

RELEVANCE

This finally brings us to the relevance thesis. I understand this to be the epistemological alignment on the basis of certain ontological assumptions, or ontography, which I would like to describe in the following. In attempts at classification, the conventional order postulated for the "major" sciences is mathematics, physics, chemistry, biology, psychology, and so on. Entirely different meanings can be intended with such an order. Some authors (especially earlier ones) understand it as a systematization of the objects of study; others see the focus on regularities and reliability of the statements made in the specific field; and for some, it is a simple hierarchy of scientific character. Although such a classification does not comprehensively reflect all the essential connections between the sciences, one may conclude that mathematics is present in all of the subsequent disciplines, yet neither physics nor chemistry nor any of the contents of the subsequent disciplines are present in mathematics. This fundamental asymmetry is also the reason why the reduction of chemistry to physics is normally discussed but not the reduction of physics to chemistry.

A classification system of disciplines as a framework in this regard becomes interesting particularly with respect to the various transitional areas as this is where the "major" philosophical topics lurk. I make use of the term "fulguration" in the sense that Konrad Lorenz (1903–1989) suggests. Lorenz rejects the term "emergence" as he feels the notion of "surfacing" from a fundamentally familiar, and therefore predictable, situation poorly suits his purposes because he seeks to describe the formation of that which is new and completely unexpected. Yet, fulguration is nonetheless in essence an expression of emergence theory. Lorenz describes in a decidedly naturalistic manner the evolutionary steps of nature—for example, in the origin of new species but also of those lying between the areas of reality or "layers of being" (inorganic, organic, mental, psychic).[11] The origin of the world then corresponds to a fulguration resulting in the formation of the material substrate for everything else, the origin of life from nonlife is a second fulguration, and the formation of sentient life forms and that of the mind are further fulgurations. If one then assigns the layers of being to the corresponding sciences, then, almost without exception, the order physics, biology, psychology described above will result (mathematics as a formal or structural science drops out of this contemplation). Connected with the first fulguration and, therefore, with physics are the crucial questions of natural philosophy about the origin and structure of the world (the cosmos), that is, cosmology and cosmogony. Problems of space-time (theory of relativity) and the structure of matter (radioactivity, elementary particles, quantum mechanics) also fall within this realm. Yet, where in this model (or similar models) does chemistry belong? Chemists apparently work in a field of science that seems scarcely relevant to the overall human orientation and worldview, despite the fact that the setting for their actions is undoubtedly the mesocosm. The next major topics, the epistemological *big points*, are the concepts of life and evolution which are "occupied" by biology, followed by the emergence of spirit and mind (or the mind-body problem). Localizing chemical phenomena in the mesocosm cannot be regarded as the reason for philosophical obliviousness to them. Now, the model used here certainly cannot claim to have broad acceptance. I am using it here primarily to illustrate the essence of the relevance thesis. The fulguration or emergence metaphor may be applicable to smaller, localized phenomena in chemistry, such as the formation of compounds with characteristics that cannot be predicted from knowledge of the constituents. Any way you look at it, the *major* topics, at least for the moment, are all occupied by the other sciences and, therefore, these sciences have had their "own" philosophy departments for a long time and are thus more attractive for analysis. In any case, in this way we arrive at an approach to explaining the philosophical undervaluation of the science of substances. This approach also provides a more plausible way to explain the widespread acceptance of the philosophical relevance of neighboring disciplines than do the other theses.

What could a strategy for the philosophical revaluation of chemistry look like? The simple postulate that a science would be relevant for philosophy when it (or better, its applications) has achieved societal (economic and industrial) significance seems to me to be insufficient. At least three other ways of improving chemistry's philosophical status are conceivable. One such strategy, admittedly a not particularly likable and also rather naïve one, would be the "hostile takeover" of the life sciences. If it were possible to convincingly and comprehensively place the phenomenon of life on a chemical foundation, then biology would lose its philosophical fascination and chemistry would be enhanced accordingly. Efforts in such a direction have been present for some time and broad areas of modern laboratory biology are basically, at least in their methodology, now no different from chemistry, although they bear the name molecular *biology*. Yet, whether the capabilities of the sciences of substances will ever suffice to explain life entirely remains very doubtful. Konrad Lorenz addressed the subject of non-reducibility of living systems to inorganic matter, noting with skepticism:

> Yet the same applies in a like manner to the man-made machines that for that very reason provide a good illustration for the nature of the non-reducibility intended here. If one considers only their current physical framework of action, then they can be completely analyzed, down to the ideal proof of a successful analysis, down to complete feasibility of synthesis, namely practical producibility. Yet if one considers their historical, telenomic process of having become organs of homo sapiens, then, in attempting to explain their being this way and not another, one comes up against exactly the same non-rationalizable rest as with living systems. (Lorenz 1977, 54)

Second, the reverse and optimistic interpretation of Schummer's *dematerialization* thesis offers important perspectives: if it is possible to construct a substance-related philosophy, then chemistry as the reference science may be expected to automatically become more the focus of discussion in the philosophy of science. In any case, it is completely correct, actually even self-evident, with respect to the sciences of substances to give serious philosophical consideration to the significance of the concept of substance and all aspects related to it (see the discussion of self-descriptions above). Schummer himself importantly contributed to such a substance-related philosophy of chemistry, and in many aspects, I base my own efforts on these contributions. The most monumental epistemological and ideological questions will probably be addressed to a lesser extent in the philosophy of chemistry, but certainly questions in the "middle range" of practical human experience, and in this sense, the postulate of a scientific discipline with societal relevance should not be dismissed out of hand. These include such questions as what precisely *metabolism* is and what influence its natural and technical forms have on the environment. Roald Hoffmann, Nobel Prize winner in 1981, provides a clear message in this regard: "Chemistry is the truly anthropic science—our molecules can heal, and they can hurt, for they are on the scale of the molecules in our bodies" (Hoffmann 2007, 333). Because he also uses microphysical language, such a comment leaves open whether the mesocosmic aspects mentioned are not best considered part of the life sciences.

Third, a culturalistic persuasion such as that advocated by Peter Janich and his followers leads to the opportunity (indeed the necessity) for operationalist reconstruction of chemistry as well as other areas. Culturalism follows the view that regularities are not discovered in or with nature but that they are quasi-constructed by researchers. This means that not only is the respective theoretical content emphasized and interpreted in the philosophy of science but also that their practical background (the *practices*) determines our concepts. It is true that scientific results do not appear out of nowhere. The philosophical reconstruction of practices is a valuable approach on the path to a philosophy of chemistry because its reference discipline is an experimental science par excellence. However, even culturalists must first specify what they want to understand as "chemistry." Aside from the purely reconstructive work, which would also come into question for carpentry or goldsmithing, another way must be found to determine where the actual philosophical topics lie. A system that sees chemical concepts fundamentally based on the living environment and that refers only to purposes also risks ignoring the history and reality of chemical activities, for chemistry creates not only knowledge for dominion but also knowledge for orientation.

One perhaps trivial aspect should not be neglected anyway. Only those who know at least some basics of that science can recognize whatever might be interesting in chemistry from a philosophical perspective. There is no metachemistry without chemistry.

CONCLUSION

After about thirty years of attempts to explore the foundations of chemistry, the position of the scientific discipline which is devoted to substances, their changes, and dispositions in philosophy is still pitiful. Although by far not all objects of chemical investigations have a substantial character, substances are still the main point of interest in chemistry (type 1 definition). Making substances and investigating their properties and relations might seem to be not as fascinating as string theories, relativity, quantum phenomena, the big bang theory, evolution, and mind-body problems. We still must conclude that (albeit alleged) reducibility, the neglect of the substance concept (type 3 definition), and the curiously hidden relevance are the most important reasons for the philosophical neglect of chemistry. Evidently, the official portrayal and the self-perception of chemists still play an unfortunate and obstructive role in that situation—although the existing pluralism is not necessarily something harmful. To take up a notion invented by Rein Vihalemm: not fitting the ideal picture of a φ-science is by no means a philosophical handicap for chemistry. Hence, we should keep on looking for the more interesting aspects in the non-φ-science parts.

NOTES

1. I thank John Grossman for the translation of most parts of the text, particularly the citations, and the two anonymous reviewers who helped to improve the text. I am grateful to Ave Mets for her engagement regarding the present publication and our joint pluralism project, particularly for her patience regarding the latter. With respect to that project, my thanks go also to Endla Lõhkivi for hosting me in Tartu and Apostolos Gerontas for his accompaniment.

2. Cf. the former version of my thoughts in Ruthenberg (1996).

3. Other facets of his position appear not as convincing to me—for example, his categoric criticism of "Kantianism." As far as I am concerned, Kant's views with respect to chemistry are much more valuable than the standard interpretations assume. Compare, as to this aspect, Ruthenberg (2022). Additionally, it is not easy to simply neglect Kant's insights on the "Ding-an-sich." As to that aspect, see Hasok Chang's critical remarks in Chang (2022).

4. See the contribution of Ave Mets to the present volume and Ruthenberg and Mets (2020).

5. For a discussion of chemical underdetermination, see Ruthenberg (2020a).

6. Other approaches can be found, for example, in Ruthenberg and van Brakel (2008).

7. In his excellent and elaborate essay on the philosophical neglect of chemistry, van Brakel particularly investigates its history (van Brakel 1999). The same paper contains an enlightening discussion of the peculiarities of chemistry.

8. Unfortunately, *Hyle* has been stopped recently (due to the lack of qualitative manuscripts, as the founder and editor Joachim Schummer complained), and the *Foundations of Chemistry* seems to be transformed into a journal of theoretical chemistry, as it were.

9. Compare, among others, Baird, Scerri, and Lee (2006); Earley (2003); Hendry, Needham, and Woody (2012); Ruthenberg and van Brakel (2008); Scerri and Ghibaudi (2020); Scerri and McIntyre (2015).

10. As to the extraordinary role of Ostwald, see Ruthenberg (2022). The list of scholars mentioned here, of course, cannot claim completeness.

11. See Lorenz (1977, especially ch. II). Lorenz was an exceptional natural scientist (won the Nobel Prize in physiology in 1973), yet also an avowed Nazi Party member from 1938 on.

REFERENCES

Baird, Davis, Eric Scerri, and Lee McIntyre, eds. 2006. *Philosophy of Chemistry: Synthesis of a New Discipline*. Dordrecht: Springer. https://doi.org/10.1007/1-4020-3261-7.

Cammann, Karl. 1992. "Analytical Chemistry—Today's Definition and Interpretation." *Fresenius' Journal of Analytical Chemistry* 343: 812–13. https://doi.org/10.1007/BF00328560.

Chang, Hasok. 2022. *Realism for Realistic People: A New Pragmatist Philosophy of Science*. Cambridge: Cambridge University Press. https://doi.org/10.1017/9781108635738.

Earley, Joseph E., Sr., ed. 2003. Chemical Explanation: Characteristics, Development, Autonomy. *Annals of the New York Academy of Sciences* 988 (1). New York: The New York Academy of Sciences.

Hendry, Robin, Paul Needham, and Andrea Woody, eds. 2012. *Handbook of the Philosophy of Science Volume 6: Philosophy of Chemistry*. Amsterdam: Elsevier.

Hoffmann, Roald. 2007. "What Might Philosophy of Science Look Like If Chemists Built It?" *Synthese* 155 (3): 321–36. https://doi.org/10.1007/s11229-006-9118-9.

Janich, Peter. 1980. "Chemie." In *Enzyklopädie Philosophie und Wissenschaftstheorie,* edited by Jürgen Mittelstraß. Mannheim, etc.: Bibliographisches Institut, 389–90.

Janich, Peter. 1992. *Grenzen der Naturwissenschaft*. München: Verlag C. H. Beck.

Liegener, Christoph, and Guiseppe Del Re. 1987. "Chemistry vs. Physics, the Reduction Myth, and the Unity of Science." *Zeitschrift für allgemeine Wissenschaftstheorie* 18: 165–74. https://doi.org/10.1007/BF01801083.

Lockemann, Georg. 1950. *Geschichte der Chemie*. Berlin: Walter de Gruyter & Co.

Lorenz, Konrad. 1977. *Die Rückseite des Spiegels*. München: Deutscher Taschenbuch Verlag.

Ostwald, Wilhelm. 1907. *Prinzipien der Chemie. Eine Einleitung in alle chemischen Lehrbücher*. Leipzig: Akademische Verlagsgesellschaft m.b.H.

Plath, Peter Jörg. 1990. "Chemie." In *Europäische Enzyklopädie zu Philosophie und Wissenschaften*, edited by Hans Jörg Sandkühler. Hamburg: Meiner.

Primas, Hans. 1985. "Kann Chemie auf Physik reduziert werden? Erster Teil: Das Molekulare Programm." *Chemie in unserer Zeit* 19 (4): 109–19. https://doi.org/10 .1002/ciuz.19850190402.

Römpp, Hermann. 1975. *Römpps Chemie-Lexikon.* 7. Aufl. Stuttgart: Franckh'sche Verlagshandlung.

Ruthenberg, Klaus. 1996. "Warum ist die Chemie ein Stiefkind der Philosophie?" In *Philosophie der Chemie: Bestandsaufnahme und Ausblick,* edited by Nikos Psarros, Klaus Ruthenberg, and Joachim Schummer, 27–35. Würzburg: Königshausen & Neumann.

Ruthenberg, Klaus. 2020a. "'Caught in the Amber': A Sketch of Chemical Underdetermination." In *Uncertainty in Pharmacology: Epistemology, Methods, and Decisions,* edited by Adam LaCaze and Barbara Osimani, 173–84. Cham: Springer Nature. https://doi.org/10.1007/978-3-030-29179-2_8.

Ruthenberg, Klaus. 2020b. "Making Elements." In *What Is a Chemical Element?* edited by Eric Scerri and Elena Ghibaudi, 204–24. Oxford: Oxford University Press.

Ruthenberg, Klaus. 2022. *Chemiephilosophie.* Berlin and Boston: De Gruyter. https:// doi.org/10.1515/9783110740493.

Ruthenberg, Klaus, and Ave Mets. 2020. "Chemistry Is Pluralistic." *Foundations of Chemistry* 22: 403–19. https://doi.org/10.1007/s10698-020-09378-0.

Ruthenberg, Klaus, and Jaap van Brakel, eds. 2008. *Stuff: The Nature of Chemical Substances.* Würzburg: Königshausen & Neumann.

Scerri, Eric, and Lee McIntyre, eds. 2015. *Philosophy of Chemistry: Growth of a New Discipline.* Dordrecht: Springer. https://doi.org/10.1007/978-94-017-9364-3.

Schmidt, Max. 1967. *Anorganische Chemie,* Teil 1. Mannheim: Bibliographisches Institut. https://doi.org/10.1515/9783112320334.

Schummer, Joachim. 1996. "Philosophie der Stoffe, Bestandsaufnahme und Ausblicke: Von der philosophischen Entstofflichung der Welt zur ökologischen Relevanz einer Philosophie der Stoffe." In *Philosophie der Chemie: Bestandsaufnahme und Ausblick,* edited by Nikos Psarros, Klaus Ruthenberg, and Joachim Schummer, 143–64. Würzburg: Königshausen & Neumann.

Ströker, Elisabeth. 1967. *Denkwege der Chemie: Elemente ihrer Wissenschaftstheorie.* Freiburg and München: Verlag Karl Alber.

Ströker, Elisabeth.1968. "Element und Verbindung. Zur Wissenschaftsgeschichte zweier chemischer Grundbegriffe." *Angewandte Chemie* 80 (18): 747–53. https:// doi.org/10.1002/ange.19680801807.

Van Brakel, Jaap. 1999. "On the Neglect of the Philosophy of Chemistry." *Foundations of Chemistry* 1: 111–74. https://doi.org/10.1023/A:1009936404830.

Van Brakel, Jaap. 2014. "Philosophy of Science and Philosophy of Chemistry." *HYLE—International Journal for Philosophy of Chemistry* 20: 11–57.

Vihalemm, Rein. 2012. "Practical Realism: Against Standard Scientific Realism and Anti-Realism." *Studia Philosophica Estonica* 5(2): 7–22. https://doi.org/10.12697 /spe.2012.5.2.02.

Chapter 9

What Does Chemistry Do?

Alexander Pechenkin and Apostolos K. Gerontas

In his PhD thesis and subsequent papers, Rein Vihalemm discussed the problem of the specificity of chemistry as a natural science. While in the 1970s, he discussed this issue by invoking the concept of the structural levels of matter, in the 1980s, he shifted to approaching this problem as epistemological, discussing the frontier between physics and chemistry and referring to the structure of scientific research. Vihalemm emphasized the dual nature of chemistry, portraying it as a φ-science hosting the constructive-hypothetico-deductive method (similar to ideal physics). Simultaneously, it is a classifying and systematizing science (similar to natural history in the past). This chapter supports Vihalemm's position by presenting the example of the discovery and description of the Belousov–Zhabotinsky reaction—a reaction connected with the history of microbiology and the history of chemistry as a classifying and systematizing discipline, akin to natural history. This very history, as we demonstrate, is linked to the evolution of nonlinear nonequilibrium thermodynamics, formulated by Ilya Prigogine and his coauthors. Therefore, it serves as an example of φ-science, as described in Vihalemm's works.

IS CHEMISTRY REDUCIBLE TO PHYSICS? AN ONTOLOGICAL APPROACH

One of the main topics that occupied the mind of Rein Vihalemm as a philosopher of science was the issue of demarcation between chemistry and physics. How does chemistry maintain its independence among the natural sciences despite the relevant successes of physics in explaining chemical phenomena? While working on his PhD thesis, he took the position of the structural levels of matter, going back to the Marxist classic Friedrich Engels, then revered in the USSR. He followed his supervisor Bonifaty Mikhailovich Kedrov, a well-known Soviet philosopher and a specialist in the history and philosophy of chemistry.[1]

Building on Engels's work, Kedrov wrote about the forms of motion of matter. Engels's scheme contained the following sequence of forms of motion: mechanical (the motion of macroscopic bodies), physical (the motion of molecules), chemical (the motion of atoms), and biological (the processes involving proteins and presupposing metabolism). Soviet philosophers of the time, working on issues of scientific disciplinary *hierarchy* (a fundamental concept in Engels's epistemology), would most certainly have the following quote of his in mind:

> Motion in the most general sense conceived as the mode of existence, the inherent attribute of matter, comprehends all changes and processes occurring in the universe, from mere change of place right up to thinking. The investigation of the nature of motion had, as a matter of course, to start from the lowest, simplest forms of this motion and to learn to grasp these before it could achieve anything in the way of explanation of the higher and more complicated forms. Hence, in the historical evolution of the natural sciences, we see how first of all the theory of simplest change of place, the mechanics of heavenly bodies and terrestrial masses, was developed; it was followed by the theory of molecular motion, physics, and immediately afterward, almost alongside it and in some places in advance of it, the science of the motion of atoms, chemistry. Only after these different branches of the knowledge of the forms of motion governing non-living nature had attained a high degree of development could the explanation of the processes of motion represented by the life process be successfully tackled. (Engels 1940, 85)

Engels insisted on hierarchy. Zbigniew A. Jordan writes,

> Engels made constant use of the metaphysical insight that the higher level of existence emerges from and has its roots in the lower, that the higher level constitutes a new order of being with its own irreducible laws and that this process of evolutionary advance is governed by laws of development which reflect basic properties of "matter in motion as a whole," (Jordan 1967, 167)

Already in the second half of the nineteenth century, physics posed a challenge to Engels's scheme of the forms of motion. Electricity and magnetism had no place there. Engels discussed these phenomena but did not incorporate them into his views.

The rise and development of quantum physics necessitated a fundamental reconstruction of the Engelsian scheme. Already in 1929, one of the most significant physicists of the twentieth century, Paul A. M. Dirac, proclaimed: "The underlying physical laws necessary for the mathematical theory of a large part of physics and the whole of chemistry are thus completely known [from quantum mechanics]" (Dirac 1929, 714). In the same year, Niels Bohr maintained a similar attitude (less of a position), concerning the reduction of chemistry into physics in his works *The Atomic Theory and the Fundamental Principles Underlying the Description of Nature* (Bohr [1929] 1985), as well as his introduction in the *Atomic Theory and the Description of Nature* (Bohr [1929] 1934).[2] In the 1930s and later, theoretical chemists Henry Eyring, John Walter, and George E. Kimball proclaimed almost the same: "In so far as quantum mechanics is correct, chemical questions are problems in applied mathematics" (Eyring, Walter, and Kimball 1936, iii).

Kedrov attempted to accommodate the fact of the development of quantum chemistry while preserving Engels's scheme of the forms of motion. He maintained the idea of hierarchy and distinguished between two forms of motion described by physics: microphysical (the motion of electrons and other fundamental particles) and molecular. Kedrov then positions chemical processes between the two.

Vihalemm's conceptual innovation consisted of applying the concept of *structure* in the analysis of the relationship between chemistry and quantum mechanics. Following some of the specialists in the philosophy of science (Moscow, Leningrad [now St. Petersburg], and Tallinn), Vihalemm considered "structure" as a capital methodological category which allowed the philosophers to discuss the modern development of science linked to the conceptual innovations provided by quantum mechanics and by nonclassical physics in general.

In contrast to the concept of the forms of motion (which can be traced back to Hegel's *Naturphilosophie* and bears the stamp of scholasticism), the concept of structure was an element of a "scientific" language. Furthermore, the use of this concept paved the way for the use of several related concepts, such as "element," "component," "invariance," "symmetry," and "the structural levels of matter." Vihalemm followed in this Lembit Valt, a philosopher from Tallinn, Estonia, who took the concept of the structural levels of matter under consideration and emphasized the subordination and coordination of structures. Valt (1963) highlighted the idea of the relativity of structural levels.

During the 1960s, a conceptual shift occurred in Soviet philosophy—and an important one. In the Soviet intellectual environment—where "philosophy" essentially stood for "Marxist-Leninist-Stalinist philosophy" (and was not merely referring to philosophy, but to a holistic ideology about life)—the shift from the forms of motion to "structure" and "the structural level of matter" had broader repercussions and meaning, which, however, fall outside the scope of this chapter.

On the epistemological level, Soviet scientists of the era had to keep in mind the (communist) philosophical attack on the Copenhagen interpretation of quantum mechanics, which had occurred in the immediate post–Second World War period. Chemists had to remember the Union-wide conference on the theory of chemical structure, which took place in 1951. This conference, in principle, disavowed the quantum chemical theory of resonance, elaborated by the American chemists Linus Pauling and George W. Wheland in the early years of quantum chemistry. Together with the theory of resonance, the very idea of quantum chemistry was attacked because it was constructed according to the "metaphysics of reductionism"—it suggested reducing chemistry to quantum mechanics. Admittedly, the situation was ambiguous: while quantum chemistry was criticized by the ideologists and by some of the chemist-experimentalists, it was not prohibited (see Graham 1987; Pechenkin 1995). It should be noted, however, that among the critics of resonance was also Kedrov, Vihalemm's supervisor.

The new Soviet structuralism allowed Vihalemm to welcome quantum chemistry as one of the achievements of early twentieth-century science. Vihalemm joined the physicist Mikhail Veselov, a specialist in the quantum theory of chemical systems from St. Petersburg. According to Veselov, "quantum mechanics basically has the potential to correctly explain all phenomena occurring in the electron shells of all systems, no matter the number of atoms entering the system, and therefore it is a theory both of physical and chemical *elementary phenomena*" (Veselov 1962, 213; Vihalemm 2021, 115).

Nevertheless, the conception of the structure of matter remained linked to the idea of a hierarchy of the layers of matter in the Soviet philosophical tradition. As Vihalemm saw it in 1965:

> Since the subject of atomic-electronic physics is the motion of electrons in the force field of a nucleus, but molecules do not form *immediately* from nuclei and electrons in chemical processes, then the possibility of reducing chemistry to physics always *remains* a mere *possibility*. As soon as we start realizing this possibility, we no longer stay in the sphere of physics, but proceed to a science about a relatively higher structure of matter—chemistry. This structural level is governed by its own objective laws which express the qualitative facet of this level which is inaccessible to the laws of a lower structure, which can only explain this level from a quantitative aspect. *Both sciences have different cognitive resources and must consider themselves in consonance.* (Vihalemm 2021, 122)

Still, Soviet structuralism at least introduced the problem of the quantum explanation of the phenomenon of chemical bonding into the debate. While Engels and Kedrov insisted that there exists a frontier separating physics and chemistry, Vihalemm could not help but wonder.

As the problem of the quantum theory of the elementary act of a chemical reaction was becoming topical in chemical kinetics, Vihalemm wrote:

> Here the task is set perfectly correctly—a new structure requires a new theory. If such a theory will be elaborated, then apparently it will be the core of the chemical theory. (Since the core of the *chemical theory* cannot be *chemical*: chemistry remains in the "upper layer," while physics is in the "nether layer.") The results obtained with the help of this theory will, however, belong in chemistry and acquire its new qualitative substance *on the basis of chemistry.* (Vihalemm 2021, 120)

Vihalemm thus maintained an ambivalent position: a chemical theory had physics at its "core"—but was also characterized by concepts and specificities that were strictly "chemical." This position could be seen as a form of *ontological pluralism*, of the type that is better described by Lombardi:

> Once the epistemological irreducibility of chemistry to physics is admitted, the ontological priority of the physical world turns out to be a mere metaphysical prejudice. From the pluralist viewpoint, concepts like bonding, molecular shape and orbital refer to entities belonging to the chemical ontology, which only depends on the theory that constitutes it. Chemical entities do not owe their existence to an ontologically more fundamental level of reality, but to the fact that they are described by theories whose immense predictive and creative power cannot be ignored. (Lombardi 2015, 23)[3]

It should be noted, however, that Vihalemm never articulated his ontological pluralism. Hence, our identification of his position as such should be treated as mildly speculative.

The solidification of quantum chemistry as an in-between of chemistry and physics happened with, and through, a series of ontological discussions. The discussions were partially generated by the very methodology of quantum chemistry: the method of "atoms in molecules" (Bader 1990) necessarily raised the question of whether the atom is an element of the molecule, or whether the method merely served as an effective calculative tool. Shant Shahbazian (2013; 2014) treated this problem as ontological: "Are there really atoms in molecule?"

THE SPECIFICITY OF CHEMISTRY AS AN EPISTEMOLOGICAL PROBLEM

The effective abandonment of Engels's concept of motion and the adoption of structure brought Vihalemm's work closer to contemporary Western thought.

Subsequently, working in the tradition of Western philosophy of science, Vihalemm discussed the relationship between physics and chemistry as an epistemological issue (in the 1990s and later). However, his epistemology was rather specific. He put forward the philosophy of *practical realism*.

According to Vihalemm, chemistry differs from physics in the kind of practical activity behind their respective concepts. Whereas physics tends to build rigorous predictive schemes, chemistry tends to classify and put forward the operational concepts that develop along with the development of chemical practice. Physics tends to operate with mathematical schemes that describe the whole world at once. Chemistry, on the other hand, offers theoretical generalizations that are more local and more practical (see Vihalemm 1999, 85–88), and can handle experimental systems, even before any theoretical generalizations are available.

In his words:

> Modern chemistry is a mixture of constructive hypothetico-deductive inquiry . . . and classifying-historico-descriptive inquiry. . . . If pure φ-science can really be defined by means of the laws of nature, then chemistry has to be defined through substance (or stuff), and only thereafter as a research field that studies how and to what extent substances can be treated φ-scientifically from the viewpoint of the laws of nature. (Vihalemm 2007, 231–32)

Or:

> [T]here are no subjects or objects of cognition that were "ready-made" or "given" by nature itself; both subjects and objects have a historico-cultural character. Nevertheless, we can distinguish between the objects of *φ-science* and the objects of *non-φ-science* (or natural history). The former ("free falling bodies," "electric current," "light-rays," and the like) are constructed in scientific practice whereas the latter (plants, minerals, animals, and the like) are "given" to the researcher, in some way or another, by some kind of pre-scientific (or non-scientific) practices. (Vihalemm 2013, 365)

Chemistry, then, has a "dual nature": on the one hand it is a φ-science, constructed according to a hypothetico-deductive standard of knowledge, but on the other hand, it focuses on the classification and systematization of natural phenomena, tracing their evolution over historical periods and in the course of human activity. We would therefore expect the majority of the chemical disciplinary processes to contain, as traces, elements of both natures of the discipline.

The presentation of theoretical knowledge from the perspective of its hypothetico-deductive structure is a message characteristic of the philosophy of science (Duhem, Hempel, Popper, Reichenbach, Nagel, and others). Sometimes, this presentation is established as standard. Vihalemm wrote about the constructive-hypothetico-deductive structure of knowledge in physics and, partially, in chemistry. By adding the predicate "constructive," he emphasized that he meant not only the logical structure of theoretical knowledge; he characterized knowledge from the perspective of its production as a historic and cultural phenomenon. Thus, he emphasized the significance of mental experiments and idealizations in scientific research. Vihalemm proposed the periodic law as an example of a law based on the empirical classification of substances, combined with an account of the historical development of matter in the universe.

In the following, we will focus on the discovery of the Belousov–Zhabotinsky reaction and its subsequent explanation within nonlinear nonequilibrium thermodynamics, developed by Prigogine and his coauthors. The Belousov–Zhabotinsky reaction arose within the classifying and experimental research typical of chemistry. In turn, Prigogine's theory has been constructed in the style of a constructive-hypothetico-deductive discipline: it arose within the trend to extrapolate classical thermodynamics from and over the observable phenomena. Nonlinear nonequilibrium thermodynamics is a good example of what Vihalemm called φ-science, and the accommodation of the Belousov–Zhabotinsky reaction in its theoretical frame is a good example of the construction of chemical knowledge.

THE BELOUSOV–ZHABOTINSKY REACTION WITHIN THE FRAMEWORK OF THE CLASSIFICATION AND SYSTEMATIZATION OF NATURAL PHENOMENA

The dramatic history of Belousov's discovery has been outlined in several journal issues and a book (Pechenkin 2018). In 1951, Boris Belousov discovered a homogeneous reaction associated with a periodic change of color of an entire reaction mixture from colorless to yellow, then back to colorless, and so on, thus defying what counted as common knowledge in the scientific community that such reactions occurred only in heterogeneous systems. This reaction was the oxidation of citric acid by bromate ion, a standard oxidant.

Due to his background in chemical warfare research, Belousov's interests extended to biochemistry, a field with which he had substantial experience. In 1950, he attempted to model catalysis in the Krebs cycle, utilizing the metal ion cerium instead of a protein-bound metal ion common in the enzymes of living cells. Normally, the oxidation reaction of citric acid by bromate ion is very slow. The rate of oxidation is increased if cerium cation 3+ is used as a transmitter of oxidation. Bromate ion oxidizes Ce^{3+} to Ce^{4+}. In turn, Ce^{4+} oxidizes citric acid, and cerium is reduced to Ce^{3+}. Since the solution containing Ce^{3+} is colorless and the one containing Ce^{4+} is yellow, the reaction undergoes a periodic change of color. The system, thus, is a nonlinear chemical oscillator, a category of reactions that has hence received special importance in thermodynamics.

In his posthumously published paper, Belousov noted that the peculiar behavior of citric acid in the presence of some oxidants lies "at the foundation of the periodic reaction"—he saw an analogy in the function of the Krebs cycle and his discovered reaction and called his reaction a "cycle." As Winfree noted, "[t]he Krebs cycle is called a 'cycle' not because it oscillates in time, but just because the reaction sequence leads in a circle, much as any biogeochemical cycle" (Winfree 1984, 661), hence we may assume that, for Belousov, the most apparent characteristic of his reaction was secondary for its classification: what mattered was the mechanism.

Belousov's original paper was rejected by authoritative scientific journals, and it was only published posthumously (Belousov 1981). He succeeded in publishing a small abstract of his main paper in a less reputable and non-peer-reviewed journal (Belousov 1958/1959). The reasons behind the rejection were rather plain: Belousov presented his results without being able to offer a theoretical justification that would satisfy the chemical orthodoxy. His claim that his observed results originated in a real homogeneous system was questionable, as was his heuristic strategy of building up his experiment on an analogy, and without theoretical support. Quite importantly, without any theory underlying his results, Belousov's reaction remained a strictly local peculiarity: it did not align with any recognized domain of chemical theory and resisted generalization.

It should be noted, however, that the Krebs cycle itself is a rather clear example of chemistry's inherent tendency to record and classify. It was established in the chemical canon through the gradual discovery of its components, completing the puzzle on the part of the scientific community long before chemistry, as a φ-science, provided theoretical justifications for the cycle and the involved components. One might argue that the Krebs cycle was "given" to the researchers through some kind of pre-(φ-)chemical practices. Belousov was, above all, a practitioner of chemistry in its experimental, recording, and classifying incarnation. Building his experimental setup upon an analogy, recording his observations, and outlining a reaction mechanism before acquiring theoretical justifications for this mechanism were all parts of the non-φ-science tradition that he served. Soon enough, his reaction changed hands and, in the process, it found a place in the φ-science domains of chemistry.

NONLINEAR NONEQUILIBRIUM THERMODYNAMICS:
AN EXAMPLE OF Φ-SCIENCE

The Belousov–Zhabotinsky reaction was discovered and formulated in the course of classifying-historico-descriptive inquiry. Its theoretical explanation was provided in the course of constructive-hypothetico-deductive inquiry conducted by Prigogine and his collaborators (also, another explanation, rooted in the framework of the theory of nonlinear oscillations, was constructed by Anatol Zhabotinsky [1974]).

In its classical form, thermodynamics covers the processes proceeding in an equilibrium or near-equilibrium state. In the second half of the twentieth century, the problem of how to expand thermodynamics to nonequilibrium processes (the majority of natural processes) became the main problem of thermodynamics. Thermodynamics was thus expanded almost simultaneously toward several different frontiers; for example, by invoking the concept of physical kinetics and hence the concept of statistical physics (see, e.g., the 1972 book by Yurii Rumer and Moisei Ryvkin[4]). Ilya Prigogine and his collaborators followed the tradition of the Brussels school founded by Théophile de Donder, who preferred to develop the conceptual resources of classical thermodynamics.

The 1971 monograph by Prigogine and Glansdorff represented the continued development of nonlinear nonequilibrium thermodynamics, which became the basis of the explanation of the Belousov–Zhabotinsky reaction and similar chemical phenomena. "The task addressed in this monograph," a review of the book says,

> is to extend the method of thermodynamics to fluid and chemical dynamics. A unified approach to nonlinear irreversible phenomena is sought which will be applicable from equilibrium to turbulence. Stability is the principal process discussed. Conventional entropy is the key variable. Starting with the basic conservation equations for fluid systems, the authors construct volume integrals which are related both to the stability of the fluid and to the entropy production rate. (Malkus 1972, 400)

In the Introduction, Glansdorff and Prigogine formulate the following definition of the dissipative structure:

> From the macroscopic point of view it is necessary to distinguish between two types of structure:
>
> (a)   equilibrium structures;
> (b)   dissipative structures.

Equilibrium structures may be formed and maintained through *reversible* transformations implying no appreciable deviation from equilibrium. A crystal is a
typical example of an equilibrium structure. Dissipative structures have a quite
different status: they are formed and maintained through the effect of exchange
of energy and matter in non-equilibrium conditions. The formation of cell patterns at the onset of free convection . . . is a typical example of a dissipative
structure. (Glansdorff and Prigogine 1971, 9)

Here Glansdorff and Prigogine write about the Bénard cells. In his Nobel lecture, Prigogine (1977) provided the following description of this phenomenon:

It is remarkable that this new type of behavior appears already in typical situations studied in classical hydrodynamics. The example which was first analyzed
from this point of view is the so-called "Bénard instability." Consider a hori-
zontal layer of fluid between two infinite parallel planes in a constant gravitational field, and let us maintain the lower boundary at temperature T_1 and the
higher boundary at temperature T_2 with $T_1 > T_2$. For a sufficiently large value
of the "adverse" gradient $(T_1 - T_2)/(T_1 + T_2)$, the state of rest becomes unstable
and convection starts. The entropy production is then increased as the convection provides a new mechanism of heat transport. Moreover, the state of flow,
which appears beyond the instability, is a state of organization as compared to
the state of rest. Indeed a macroscopic number of molecules have to move in a
coherent fashion over macroscopic times to realize the flow pattern. (Prigogine
1977, 267)

Nonlinear nonequilibrium thermodynamics is thus an attempt to construct
a constructive-hypothetico-deductive system that can accommodate a multitude of phenomena observed in nature and provide justifications and,
when possible, predictions of their appearance and development. As such, a
theoretical construct of this kind should be transferable from the domain of
physics to chemistry, biology (and engineering), and back, through the use of
generalized concepts and formulas. Nonlinear nonequilibrium thermodynamics is a φ-scientific system, as the one described by Vihalemm.

Prigogine and Glansdorff presented their system along these lines: Nage-
lian "bridge laws" formulated specialized practically applicable case statements that were derivable from the basic principles of their theory (Nagel
1961, ch.1, sec. II.3).[5] By constructing such auxiliary sentences, physicists
and chemists logically came to several practically significant statements that
could not be directly derived from the previously fundamental principles of
thermodynamics.

To extend thermodynamics to nonequilibrium processes, we need an
explicit expression for the entropy production," Prigogine said in his Nobel
Lecture,

Progress has been achieved along this line by supposing that even outside equilibrium entropy depends only on the same variables as at equilibrium. This is the assumption of "local" equilibrium" [an example of the "bridge law"—Authors.]. Once this assumption is accepted we obtain for P, the entropy production per unit time, [formula] (2.3) where the J_p are the rates of the various irreversible processes involved (chemical reactions, heat flow, diffusion . . .) and X_p the corresponding generalized forces (affinities, gradients of temperature, of chemical potentials . . .). This is the basic formula of macroscopic thermodynamics of irreversible processes.

At thermodynamic equilibrium we have simultaneously for all irreversible processes $J_p=0$ and $X_p=0$. It is therefore quite natural to assume, at least near equilibrium, linear homogeneous relations between flows and forces [another example of the "bridge law"]. . . . We obtain in such a way linear thermodynamics of irreversible processes. (Prigogine 1977, 265–66)

Drawing on Vihalemm's terminology, it could be argued that these bridge laws have been constructed employing concepts that are not present in the basic vocabulary of thermodynamics. These concepts are products of idealization following mental experiments, and the formulation of mathematical pronouncements, which are then used to capture and order reality.

Glansdorff and Prigogine formulate, at the beginning of their book, the hypothesis of the local equilibrium. The hypothesis is based on the idea that a system that is not currently in (so-called global) equilibrium can still be considered as consisting of subsystems that are describable using equilibrium standards and handled by equilibrium rules. In nature, most systems can be handled that way (Vilar and Rubí 2001). In contrast to the classical global equilibrium—which is considered a homogeneous state for the whole of a thermodynamic system—the local equilibrium allows theoretical pronouncements for real systems: calculating parameters for proposed subsystems that can be roughly considered in equilibrium and then creating the overall image for the system examined.

> Generally the calculation of both the entropy production and the entropy flow presupposes the method of statistical mechanics. However we shall take such situations where this calculation is possible within the framework of thermodynamics. We take under consideration the situations where the local equilibrium exists: the local entropy is such a function of the local macroscopic variables as it is in the state of equilibrium. (Glansdorff and Prigogine 1971, 29)

The assumption of the linearity of the equations relating thermodynamic flows and thermodynamic forces follows the assumption of local equilibrium. True, this assumption concerns only the thermodynamics of irreversible processes close to equilibrium. "It is quite natural to assume at least near equilibrium linear homogeneous relations between flows and forces. . . . We obtain in this way the linear thermodynamics of irreversible processes" (Prigogine 1977, 266). This thermodynamics is characterized by two prevailing results: the Onsager relations and the theorem of minimum entropy production.

As the next step (in chapters 4–9), Glansdorff and Prigogine developed the nonlinear thermodynamics of nonequilibrium states and arrived at the concept of dissipative structures. They formulated the thermodynamic theory of stability for the processes that occur with large deviations from equilibrium. Again, they constructed the bridge laws, which referred to the new key concepts (the excess of the entropy flow, the excess of the entropy production).

Glansdorff-Prigogine's book belongs to both physics and chemistry, as would be expected by a hard φ-science treatise. It is structured into three parts: (1) general theory, (2) variation technique and hydrodynamic applications, and (3) chemical processes. The third part presents the theoretical explanation of the Belousov–Zhabotinsky reaction.

To explain the Belousov–Zhabotinsky reaction, Glansdorff and Prigogine used the technique of Poincaré limit cycles (the technique of the qualitative theory of differential equations). In this process of explanation, the reaction assumes its position in a family of oscillatory processes—a position that is effectively dictated by the theoretical framework used. Glansdorff and Prigogine applied their theory of thermodynamic stability to express the stability of the Belousov–Zhabotinsky periodic process, described as the Poincaré limit cycle. One could reasonably argue that, with the accommodation of the Belousov–Zhabotinsky reaction in the theory of thermodynamics, the whole trajectory of a chemical factum had been completed: from the natural-historic description and classification to the theoretical formalization and inclusion, in the safety of φ-science structure.

CONCLUSION

While it is relatively easy to recognize the coexistence of two research traditions in modern chemistry—an "ideally scientific" and a "natural-historic"—and hence a dual nature, one should, of course, be careful. These two traditions are present in the history of all scientific disciplines and are naturally traceable in each and every one of them. Such a dual nature is easily recognizable in biology or geology and, despite the occasional pretension, is also visible in physics.

The problem of defining the specifics of a scientific discipline is a complicated one, and it is difficult (or impossible) to put forward a noncontroversial definition of any science. Every discipline is, first of all, an organizational-social-cultural unit that accommodates a multitude of domains, research programs and agendas, diverging professional interests, and, as we mentioned, (at least?) two natures or traditions. Quite often—and perhaps more often than not—a piece of scientific knowledge is generated by one agenda and methodological tradition and is completed by another. Often, a factum enters our mental world through observations and constructs belonging to one discipline but is warmly accommodated by the theory of another—either because the interdisciplinary borders have shifted or because the theoretical framework of the latter discipline is more welcoming.

Physics is also preoccupied with the problem of observation and classification (e.g., the classification of elementary particles), and via cosmology, physics is intimately connected with natural history. The existence of sociocultural (and, at least partially identical-psychological) categories such as "scientific discipline" and "research area" proves effective, as such categories can be explored historically and philosophically through "harder" emanations and traces: institutional structures such as departments, academies, professional unions, journals, conferences, bibliometric and scientometric data. Through these categories, one can engage in philosophy of science in the first place. It is only rarely, however, that a piece of scientific knowledge can seriously be attributed to one and only of these categories.

Still, following Vihalemm, one could argue that, while reality is singular and indifferent to human constructs such as disciplines, the "taste of reality" varies considerably across different disciplines, and this variation is substantial enough to lend meaning to the classification and investigation into methods and structures. The investigation on the definition of a scientific discipline is interesting, we think, raising questions about the nature of observation and the limits and definitions of the observable, the nature and the structure of a theory, and the paths to its construction, as well as the effects that human ideologies (and disciplinary expectations) have on scientific processes. As Vihalemm would (probably) say, scientific reality is not only temporal; it is also local—reality in physics differs from reality in chemistry.

NOTES

1. Kedrov contributed to the study of the history of Mendeleev's discovery of the periodic law, and he described Mendeleev's compilation of the system of chemical elements. Kedrov insisted that one should strictly distinguish between the table of chemical elements and the periodic law, which manifests itself in the table.

2. Interestingly enough, in these works, Bohr handled the issue of the reduction of chemistry to physics as solved—but the potentiality of the reduction of biology to physics as almost impossible.

3. See also Lombardi and Labarca (2004).

4. For the English translation, see Rumer and Ryvkin (1980)—Ed.

5. See also van Riel and Van Gulick (2023).

REFERENCES

Bader, Richard F. W. 1990. *Atoms in Molecules: A Quantum Theory*. Oxford: Oxford University Press. https://doi.org/10.1093/oso/9780198551683.001.0001.

Belousov, Boris P. 1958/1959. "Periodicheski deistvuyushchaya reaktsia i ee mekhanism" [Periodically Acting Reaction and Its Mechanism]. In *Sbornik referatov po radiotsionnoi meditsine, 1958* [Collection of Abstracts on Radiation Medicine, 1958], 145–47. Moscow: Medgiz.

Belousov, Boris P. 1981. "Periodicheski deistvuyushchaya reaktsia i ee mechanism" [Periodically Acting Reaction and Its Mechanism]. In *Avtovolnovye protsessy v sistemakh s diffuziei* [Autowave Processes in the Systems with Diffusion], 178–86. Gorky: Izd. Gor'kovskogo universiteta.

Bohr, Niels. [1929] 1934. *Atomic Theory and the Description of Nature*. Woodbridge, CT: Ox Bow Press.

Bohr, Niels. [1929] 1985. *The Atomic Theory and The Fundamental Principles Underlying the Description of Nature: Atomteorien og Grundprincipperne For Naturbeskrivelsen, Niels Bohr Collected Works*, volume 6, edited by Jørgen Kalckar, 219–53. North Holland: Elsevier. https://doi.org/10.1016/S1876-0503(08)70340-2.

Dirac, Paul Adrien Maurice. 1929. "Quantum Mechanics of Many-Electron Systems." *Proceedings of the Royal Society of London* 123 (792): 714–33. https://doi.org/10.1098/rspa.1929.0094.

Engels, Friedrich. 1940. *Dialectics of Nature*. Marx Engels Archive, vol. XXVII. Accessed October 2023, https://www.marxists.org/archive/marx/works/1883/don/index.htm.

Eyring, Henry, John Walter, and George Kimball. 1946. *Quantum Chemistry*. New York and London: John Wiley.

Glansdorff, Peter, and Ilya Prigogine. 1971. *Thermodynamics of Structure, Stability and Fluctuations*. New York: Wiley Interscience.

Graham, Loren R. 1987. *Science, Philosophy and Human Behavior in the Soviet Union*. New York: Columbia University Press.

Jordan, Zbigniew A. 1967. *The Origin of Dialectical Materialism*. London: Macmillan.

Lombardi, Olimpia. 2015. "The Ontological Autonomy of the Chemical World: Facing the Criticisms." In *Philosophy of Chemistry: Growth of a New Discipline*, Boston Studies in the Philosophy and History of Science, edited by Eric Scerri and Lee McIntyre, 23–38. Dordrecht: Springer. https://doi.org/10.1007/978-94-017-9364-3_3.

Lombardi, Olimpia, and Martín Labarca. 2004. "The Ontological Autonomy of the Chemical World." *Foundations of Chemistry* 7 (2): 125–48. https://doi.org/10.1007/s10698-004-0980-6.

Malkus, Willem V. R. 1972. "Review (of Glansdorff-Prigogine's Book)." *Journal of Fluid Mechanics* 53 (2): 400. https://doi.org/10.1017/S0022112072210217.

Nagel, Ernest. 1961. *The Structure of Science: Problems in the Logic of Scientific Explanation*. London: Routledge.

Pechenkin, Alexander. 1995. "Anti-Resonance Campaign in the Soviet Science." *LLULL* 18: 135–66.

Pechenkin, Alexander. 2018. *The History of Research on Chemical Periodic Processes: Springer Briefs in History of Science and Technology.* Cham: Springer Nature Switzerland. https://doi.org/10.1007/978-3-319-95108-9.

Prigogine, Ilya. 1977. "Time, Structure, and Fluctuations." Nobel Lecture, December 8, 1977, 263–85. Accessed October 2023, https://www.nobelprize.org/uploads /2018/06/prigogine-lecture.pdf.

Rumer, Yurii, and Moisei Ryvkin. 1980. *Thermodynamics, Statistical Physics and Kinetics*, translated from Russian by S. Semyonov. Moscow: Mir Publishers.

Shahbazian, Shant. 2013. "Beyond the Orthodox QTAIM: Motivations, Current Status, Prospects and Challenges." *Foundations of Chemistry* 15 (3): 287–302. https://doi.org/10.1007/s10698-012-9170-0.

Shahbazian, Shant. 2014. "Letter to the Editor: Are There 'Really' Atoms in Molecules?" *Foundations of Chemistry* 16 (1): 77–84. https://doi.org/10.1007/s10698 -013-9187-z.

Valt, Lembit. 1963. "O sootnoshenii elementov i struktury" (On the Correlation Between Elements and Structure). *Voprosy Filosofii* 5: 44–53.

Van Riel, Raphael, and Robert Van Gulick. 2023. "Scientific Reduction." In *The Stanford Encyclopedia of Philosophy* (Winter 2023 Edition), edited by Edward N. Zalta and Uri Nodelman. Accessed October 2023, https://plato.stanford.edu/ archives/win2023/entries/scientific-reduction/.

Veselov, Mikhail G. 1962. *Elementarnaia kvantovaia teoria atomov i molekul* (Elementary Quantum Theory of Atoms and Molecules). Moscow: Gos. izd. fizmat lit.

Vihalemm, Rein. 1999. "Can Chemistry Be Handled as Its Own Type of Science?" In *Ars Mutandi: Issues in Philosophy and History of Chemistry*, edited by Kostas Gavroglu and Nikos Psarros, 83–88. Leipzig: Leipziger Universitätsverlag.

Vihalemm, Rein. 2007. "Philosophy of Chemistry and the Image of Science." *Foundations of Science* 12 (3): 223–34. https://doi.org/10.1007/s10699-006-9105-0.

Vihalemm, Rein. 2013. "What Is a Scientific Concept: Some Considerations Concerning Chemistry in Practical Realist Philosophy of Science." In *The Philosophy of Chemistry: Practices, Methodologies, and Concepts*, edited by Jean-Pierre Llored, 364–84. Cambridge: Cambridge Scholars Publishing.

Vihalemm, Rein. 2021. "On the "Two-Layeredness" of Structure and the Relation of Quantum Mechanics and Chemistry." Translated by Ave Mets. *Acta Baltica Historiae et Philosophiae Scientiarum* 9 (1): 110–29. https://doi.org/10.11590/ abhps.2021.1.08.

Vilar, José M., and J. Miguel Rubí. 2001. "Thermodynamics 'Beyond' Local Equilibrium." *Proceedings of the National Academy of Sciences* 98 (20): 1191081–84. https://doi.org/10.1073/pnas.1 360398.

Winfree, Arthur T. 1984. "The Prehistory of the Belousov–Zhabotinsky Oscillator." *Journal of Chemical Education* 61 (8): 661–65. https://doi.org/10.1021/ed061p661.

Zhabotinsky, Anatoli M. 1974. *Konzentatsionnye avtokolebania* (Concentration Self-Oscillations). Moscow: Nauka.

Chapter 10

Pluralist Chemistry and the Constructive Realist Philosophy of Science

Ave Mets

Rein Vihalemm's philosophy was to a great extent influenced by his first specialization—chemistry.[1] His conception of methodological differences between scientific disciplines (e.g., Vihalemm 2016b) was directly modeled upon chemistry, which includes both physics-like aspects, which he called φ-scientific (exact scientific), and aspects more similar to natural sciences (natural history), and thus enabled Vihalemm to make this distinction between φ- and non-φ-scientific methods. In general, the φ-scientific method is constructive-hypothetico-deductive, having a mathematical core—these sciences formulate their laws in mathematical language via experimental testing, and they adapt the world to their (mathematical) cognition. They construct their object of research both theoretically and materially in laboratories. The non-φ-scientific method (here also called "natural-scientific") is classifying-descriptive-historical, looking at the phenomenon as a whole, its details, and history. While physics has often been deemed the epitome of science, at least in the popular image, but also in philosophy, Vihalemm emphasized (e.g., Vihalemm 2011a and personal communication) the importance of different scientific methodologies for their own aims. Thus, his conception of science is pluralist with respect to methodology, even if only minimally so, and this pluralism is extended to chemistry—chemistry need not and cannot be reduced to physics, but must necessarily retain both its aspects.[2]

Constructive realism in Vihalemm (e.g., 2011b) and Giere (1988, 2004) includes the constructive aspect of the exact sciences: idealized and abstract reconstruction of real-world systems and objects (e.g., point masses, frictionless surfaces, ideal gases), which are considered real to the degree that they can be materially created (i.e., in laboratory conditions). Giere's (2006a, 2006b) perspectival pluralism stems from the plurality of research devices,

219

which have different interactions with real-world systems and thus display different perspectives to, and aspects of, these, giving rise to different theoretical models. The devices are, in a trivial sense, constructed and the aspects they bring forth of the world are real.

The φ-scientific aspect of chemistry pertains to its mathematical and physical core—the physical branches of chemistry; the definition of chemical elements based on their definite atomic number, valency, and quantitative reaction properties; and the periodic system based on these. The natural-scientific aspect pertains to the research object of chemistry—stuff. Substances have multifarious properties, such as color, consistency, volatility, smell, which cannot be predicted merely from their chemical composition or deduced from mathematical models. For uncovering those, detailed experiments are needed.[3]

Recent philosophy of chemistry, which promotes the view that chemistry is inherently pluralist,[4] also links its pluralism to stuff, although for different reasons: first, the nature of chemistry as a science is that it produces a multitude of new entities and hence new constellations in the world; and second, chemical inquiry factually has a plurality of aims, and hence a plurality of research methods (not merely dual methodology).

I will consider how Vihalemm's philosophy fares with the latter kinds of plurality in chemistry, primarily that of new stuff produced. Since production is a kind of construction, I will analyze, in particular, how the "constructive" in his characterization of φ-science suits to describe chemistry. Thereby, as the productive activity indicates the centrality of practice for chemistry, and thus the relevance of its analysis, I will delve into practical realism advocated by Vihalemm. Since I deem a concrete understanding of scientific practice important, but Vihalemm has little to say about that, I will rely, for a more concrete and detailed analysis, on Ronald Giere's model-based account of science and perspectival pluralism, which Vihalemm avidly supported. This account gives an idea of how theory relates—via models and devices—to the studied part of reality, in this case, the plurality of substances. Since "substance" is a (or the) central concept in chemistry, understanding this relation is imperative for a theory of science that wants to make sense of chemistry.

STUFF AND PLURALITY

"Stuff" or "chemical substance" is the central concept of chemistry (Ruthenberg and van Brakel 2008, vii; Ruthenberg 2016, 155). The object of study in a chemical experiment is a sample of stuff, and it is the processes that lead to new objects of study that have new properties from the perspective of stuff (Schummer 1996, 96). To the notion of substance, both pluralities and pluralisms of and in chemistry are due. Plurality, Kellert et al. (2006, ix) explicate, is "a feature of the present state of inquiry." Pluralism is "a view about this state of affairs [for instance] that plurality in science possibly represents an ineliminable character of scientific inquiry and knowledge" (Kellert et al. 2006, ix).

The plurality of chemistry that primarily interests me here is that of chemical substances. Considering among the aspects of the "complexity of nature" that Kellert et al. (2006, xv) name as one of the causes necessitating pluralism about science, it underlies other pluralities, such as that of research methods and representational models. Nonetheless, stuff or substances are essential, and essentially plural, in chemistry also on their own, without the need to refer to theoretical entities (schemata, models, explanatory strategies, etc.) that they may motivate.

The plurality and multifariousness of the entities that are chemistry's research object—chemical substances—is one of the essential grounds for Vihalemm to consider chemistry among non-φ-sciences. Thus, one might say that it is not special with respect to, say, biology, which also has plural and multifarious entities—species—as the object of research (irrespective of the adopted definition of "species"). It does have important differences, however, to which Vihalemm pays no special attention: chemistry synthesizes or otherwise produces its entities, while biological science (in contrast to biotechnology) does not create species; moreover, "multiplying chemical substances is an end in itself [for organic chemistry]" (Schummer 1997b, 138; see also Simon 2012; Llanos et al. 2019; Restrepo 2022), whereas in biology, plurality is not an aim in itself—it just is, in nature.

This ontic (or ontological) plurality is theoretically represented by chemical space—the set of all possible substances and their relations in constant enlargement:

> Seen as a network, chemical space consists of the pure substances at the nodes; the relationships between the nodes are chemical reactions correlated to experimental practice. The dispositional properties of a substance include the interactions via all known and unknown chemical reactions (including reactions with as yet non-existing substances). (van Brakel 2000, 72)

The pure substances are limit reference points in the relations; for instance, conglomerates or solutions consist of several pure substances (van Brakel 2012, 196; although see Schummer 1998 for more arguments against the reality of the concept of pure substance). Chemical space is at the core of experimental chemistry and shows, inter alia, the operational way for obtaining stuff (Schummer 1996, 216–17; 1998).

Concrete elements of the chemical space are registered and described in several databases. A more general among them is the CAS Registry,[5] which for the time being contains nearly 230 million substances, descriptions of their various properties, such as reactivity, composition, density, color, taste, odor, conditions of phase transitions, as well as descriptions of producing and processing, such as separation and mixing, chemical transformation (analysis–synthesis), purifying, concentration. Describing various properties is the non-φ-scientific aspect of chemistry, according to Vihalemm (e.g., Vihalemm 2016b). What he does not take into account in his model is the producing and processing part, even though "the discovery/production and characterization of new species is actually a major task at least in chemistry" (Schummer 1997a, 108; also Schummer 1996, 229). This aspect might be accountable for Vihalemm's practical realism as practices of chemical science. Even if they sound more like technological practices, we should here keep chemical engineering apart from chemical science, as the latter has primarily epistemic aims, while the former has utility aims.

I agree with the various pluralisms about chemistry that have been put forth, such as methodological pluralism (e.g., Chang 2012; Ruthenberg and Martinez González 2017; Schummer 1996, 2015) and ontological pluralism (Lombardi and Labarca 2005). Both concern me here in a lateral sense: the former as the operational part of the chemical space—reactivities of substances, including the methods of producing them—due to the plurality of possible interactions of substances with the world and aims of creating and/or studying substances for those interactions; the latter as touching upon the realism of the chemical space as an abstract, idealized model. My central concern, however, differs from both—it is about entities themselves whose spatiotemporal existence is beyond doubt, and not about their kinds whose independence (from physics or physical entities) has been denied or doubted. So what I am doing here is rather, or at least in part, philosophy of nature—the nature, or the part of the world, that science studies and creates; this complexity of nature that underlies the need for pluralism—the plural and multifarious aspects of science that necessitate pluralism about it. It is nature inasmuch as it is the object of study of a natural science (or the natural-scientific aspect of chemistry), or in Vihalemm's sense, and inasmuch as the stuff and their relations, even though created or contrapted, are not entirely anthropogenic but rather depend on the intrinsic properties of the constituent parts. The "such" that the world is that necessitates pluralism (Kellert et al. 2006, xxii–xxiii)—the chemical world is such that it can only be grasped when studied in a variety of different ways.

REIN VIHALEMM'S MINIMAL PLURALISM AND PRACTICAL REALISM

Vihalemm's minimal pluralism, as mentioned in the introduction, is a methodological pluralism and consists of holding methodologically different sciences and their methods as legitimate in view of their own aims. According to this view, there are two kinds of methods in chemistry: φ-scientific and non-φ-scientific. The character of chemistry that I focus on here is depicted by Schummer (1996, 97) as having three essential aspects: it is (1) a classifying, (2) an experimental, and (3) a productive science. (1) Places it clearly under non-φ-sciences in Vihalemm's sense, (2) under φ-sciences, and (3) as if under φ-sciences as constructive—if production can be viewed as a kind of material construction—but with complications.[6] Let us consider the various characteristics of chemistry in comparison to Vihalemm's terms.

Since chemistry creates or constructs its own object of study—the myriad of substances (Bensaude-Vincent and Simon 2008, 99, citing Berthelot), and this is the core of chemistry I am interested in, I will start by comparing aspects and levels of construction in chemistry with those in the model "φ-science"—the host of construction in Vihalemm's account (φ-science creates its own object of study): simplicity and complexity, theoretical and material (experimental) construction. As production is clearly a practical activity, featuring certain operations and impinging on certain objects as targets of practice—"[t]he knowledge . . . and the cognisable world are formed in practice" (Vihalemm 2011b, 50). Vihalemm's practical realism, particularly its third tenet, must be considered in the discussion:

> Theoretical activity is only one aspect of science; scientific research is a practical activity whose main form is scientific experiment; the latter, in its turn, takes place in the real world itself, being a purposeful, constructive, manipulative, and material interference with nature—interference, which is, in a crucial way, theory-guided. (Vihalemm 2011b, 48)

The other most poignant features of this stance are the impossibility to get at reality outside of interactions with it and realism as materialism, that is—the reality, "carved out" via scientific material practices, is necessarily material—the dualism of theoretical and empirical knowledge is abrogated. Philosophers of chemistry, as we will see (see references in the text below), strongly endorse the materialist and practical stance.

What "Construction" Means

In φ-sciences, construction primarily means theoretical creation—the idealized, abstract objects, theoretical schemata (point masses, absence of friction, mathematical pendula, etc.). φ-sciences do not study the world or objects directly, in their complexity, but models instead—their laws pertain to idealized models. This is confirmed by Giere's account of scientific theory as a set of models, where the models central in sciences are abstract models (e.g.,

Giere 1988). This is the theoretical, and main, aspect of "creating its own object of study" in φ-sciences. In chemistry, the primary construction is not of theoretical objects or abstract models but of real material substances. Theory is not as important in chemistry as creating and describing substances (Ruthenberg 2016, 156; also Schummer 1997b). So the creation or construction of its central domain in chemistry is material, while in φ-sciences it is theoretical.

An implication of this difference concerns *simplification*. By abstracting from the confounding circumstances and idiosyncrasies of material situations, and idealizing objects into theoretical constructs, φ-sciences simplify their object of study. Ideally, they have a small set of simple and elegant mathematical descriptions of phenomena. If a more complex phenomenon needs to be described, its law (model) is built up from the simple laws. φ-sciences construct simplicity. In chemistry, the production is a construction of a plurality of real stuff and its kinds. Millions of new substances are created each year (Simon 2012, 528), and all these millions have a variety of properties and introduce innumerable new possible relations and operations into the world. Chemistry does not simplify, it complexifies.

Now one can readily object that this complexification is still accounted for with simplifications: namely, and most relevantly here, the chemical space, and that space is a construct, an idealized model, and it accounts for the stuff in an idealized way, just like physics models account for their objects—impurities and noise are abstracted away.[7] I will expand on the chemical space as a model in the next section. As for its character and role in chemistry, it is somewhat different from models in physics. It "records" what chemists have done in laboratories, and although it allows deduction to some extent, for example, about the properties and relations of substances of the same type, the non-mathematizable properties such as smell, taste, etc. must be found in real experiments; they are irreducibly material. Furthermore, as more and more substances are created, the chemical space is in constant growth, it is never complete:

> Knowledge about material properties cannot be completed, because there's no end to making new stuffs. It makes no sense therefore to refer to "intrinsic" properties of a substance "an sich," apart from real interactions. In making new substances, unpredictable relations may occur, sometimes leading to chance findings—an impurity that turns out to be something wildly new. (van Brakel 2000, 72)

Models in physics, in contrast, are completed once their relation to the material world is established satisfactorily.

Focal Activities

The focal activities in both sciences also diverge in their experimentation. In φ-science, a focal activity is deducing mathematical structures from underlying axioms—this is characteristic of Vihalemm's φ-science; in chemistry, the focal activity is materially building substances. The material experimental setups of φ-sciences are to test the mathematical theories about those idealized theoretical objects mentioned above. The construction refers to this deduction, as well as to building the (experimental) world around this theoretical-mathematical construction. Experimentation in chemistry has a different aim and meaning than in the typical exact sciences (also van Brakel 2012, 197, with reference to Schummer). As Schummer (1996, 168) notes, chemistry, in contrast to logic, mathematics, and physics, has no such axioms; it and its concepts build directly upon and are understood via experimental activity, not upon a (pre)theory or mathematical structure.

Furthermore, "[a]ny transformation of 'stuff' is first and foremost a qualitative change. No underlying quantitative description can fully grasp the 'emergent' property. Chemical systems are complex systems that cannot be reduced to their 'elements'" (van Brakel 2000, 73). Synthesis of new substances is a means to "arrive at better understanding of nature" (Berthelot 1876, cited in Bensaude-Vincent and Simon 2008, 100). And the nature of the chemical world is such—complex and irreducibly emergent—that it takes the synthesis of millions of substances to unveil it. So here, again, the φ-scientific aspect of chemistry—experimentation and experimental construction—has a rather non-φ-scientific flavor. The aim is to create new things and relations, and practical (operational) knowledge to this aim, rather than close the inquiry by the instantiation of a law.[8]

Nevertheless, I think concessions are in place here: first, also in physics, not all experiments are strictly for testing an existing explicit hunch; second, the chemical space can be seen as a (pre)theory by its role in guiding new experiments; and third, the truth of a law, according to Vihalemm, just as that of the chemical space, pertains to what can actually be done. So there is a convergence here, and thereby we see that, contrary to what Vihalemm alleges,[9] it is φ-sciences that are technological, rather than non-φ-sciences,[10] in the sense that they prescribe actions.

Manner of Constructing

Despite the different aims of chemical and φ-scientific experiments, both are experimental, creating artifacts, and this is an aspect of being technological. Chemistry creates phenomena, but also forms a "multitude of artificial entities similar to natural ones, and sharing all their properties. These artificial entities are the instantiated images of abstract laws that [chemistry] seeks to know" (Bensaude-Vincent and Simon 2008, 100). This attests to both a φ-scientific aspect—the world is adapted to our cognition, to our laws (which are prescribed by and recorded in the chemical space)—and a non-φ-scientific aspect—the detailed similarity to natural phenomena. Berthelot also believes this creation is possible in the same way as it happens in nature, from bottom up (Berthelot 1876, cited in Bensaude-Vincent and Simon 2008, 101ff.; nowadays called combinatorial chemistry, see Simon 2012). Regardless of whether or not nature does indeed build substances from bottom up, this attitude is non-φ-scientific—we must succumb to nature, know its details, adapt our cognition to it. That is, construction, which according to Vihalemm is a characteristic of φ-sciences, can in fact be a non-φ-scientific activity.

On the other hand, constructing the domain out of simple building blocks resembles the way physics accounts for complex phenomena—by assembly of simple phenomena. Theoretically, this is one of the ways in which deduction works: in exact sciences, it means, inter alia, deducing laws for complex phenomena from the laws of simple phenomena. To do this, it must hold that the complex phenomena are "analyzed" or "dissected" into simple constituent parts and reassembled, and the end result is the same as before the dissection. However, in chemistry, this is not always the case in material terms: different methods, for example, mechanical or thermodynamic ones, may be required to dissect a mixt into simple substances, and when those are mixed again, the result is different—the history of fabrication may affect a substance's properties (Schummer 1996, 178–79; van Brakel 2012, 192n5; Llored 2015). Moreover, it may affect theoretical conclusions: "Different experimental methods may lead to different conclusions about the details of molecular structure or the arrangement of nuclei and electrons" (van Brakel 2012, 218). The dissection may be necessary for classification purposes, but it destroys the object of study. A question arises about the legitimate changes through operations in substances, and the definition of pure substance (Schummer 1996, 178–79).

Stuff versus Theory

To revisit practical realism and the plurality of stuff, which are the focus of this chapter—is it even a relevant issue if it is considered as paramount to theory in a philosophical context? Is it not just material stuff lying in the dishes and containers of laboratories, a plurality of clear material, spatiotemporally discernible things, a mere fact of the matter? In a sense, it is indeed: there is an x, an amount of stuff, a sample of substance of some sort. But stating exactly which sort of substance it is is not without pitfalls. If it is a new substance, it may not be easy to determine what kind it is. It may give different results when probed by different means, for instance, based on its reactivity with stuff y, it should be classified as kind a, say, a lipid, based on spectroscopic study, it should be classified as kind b, say, a protein, and so on, and one may conclude that one is dealing with an entirely new kind of substance. However, it may turn out that the sample was contaminated and the piece of stuff is a known kind after all.[11] The ontic plurality is undeniable, but the ontological plurality concerning the kinds of substances is questionable in the sense that it takes concrete practices, operations, techniques, to determine that. And this is pervasive in chemistry: on the one hand, it is a component of the chemical space, on the other hand, "metachemistry takes into account all the technical work necessary for chemical substances to exist: they are real because they have been 'realized,' that is, synthesized or purified" (Bachelard [1940] 1968, quoted in Bensaude-Vincent 2012, 145). Thus, also according to Bachelard, chemistry is a science of effects, of action, not of facts (Bachelard [1940] 1968, quoted in Bensaude-Vincent 2012, 145). And the effects and actions are many due to substance being many.

In this section, I discussed the construction characteristic that Vihalemm assigns to φ-sciences, but it also exists in chemistry's focal non-φ-scientific mode, possessing a rather different nature there. This focus of chemistry on stuff is aptly overlooked by Vihalemm. In conclusion, his distinction between φ- and non-φ-science, even if inspired by chemistry, is oversimplified and poorly applicable to that same discipline.

RONALD GIERE'S CONSTRUCTIVE REALISM AND PERSPECTIVAL PLURALISM

Since Vihalemm's account of science and pluralism is insufficient for chemistry, I will turn to another constructivist and pluralist account that he sees as complementing his own. Giere (1988) proposes a universal account of science, asserting that its activity is constructing models, and that its theories are comprised of sets of the models. The relation of the models to the modeled real-world system hinges upon similarity—in certain aspects and to certain degrees (figure 10.1). Which aspects, and to which degree, are determined by scientists, based on the aim and purpose of the model. Real-world things have an infinite array of properties, most of which are excluded from the model. Models are defined by the terminology or mathematics of the scientific theory, yet they themselves are abstract objects (while other kinds of models exist, the abstract ones are definitive of scientific theory). Models serve to represent real-world systems in scientific reasoning, such as when calculating results of an experiment, or predicting outcomes of an action involving a phenomenon, an object, or a system.

Let us apply this idea of models to substances as objects to be modeled in chemistry. Well-known models are those of molecules, depicting different aspects of them, for example, the Lewis structure represents the bonds between atoms, but not the real geometry of a molecule, and the structural formula shows the arrangement of atoms in three dimensions (figure 10.2).

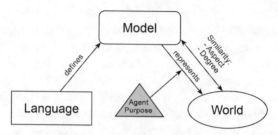

Figure 10.1   Giere's Conception of Scientific Model (Giere 1988, 83; 2004, 743).
"World" stands for the modeled part of the world (the phenomenon, system, object, attribute). "Language" stands for the terminology and mathematics included in a theory that defines the model. Source: Image courtesy of University of Chicago Press; computer graphics: Margus Evert.

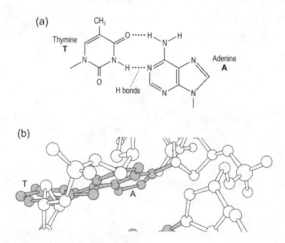

Figure 10.2   Examples of Models in Chemistry of Chemical Substances, Representing Different Aspects of the Same Object of Study (Natural World System, here DNA): (a) a two-dimensional diagram of a thymine-adenine pair and bond types included; no spatial structure shown; (b) a model showing the spatial structure of the adenine-thymine pair in the double helix, chemical elements included (distinguished with different colors).
Source: Images from Holum, John R. 1996. *Organic and Biological Chemistry*. New York: John Wiley and Sons, 336(a) and 337(b); image courtesy: (a) John Wiley and Sons, (b) Illustration close-up portion of B-DNA by Irving Geis. Image from the Irving Geis Collection, HHMI. Not to be used without permission. Computer graphics: Margus Evert.

Chemical space as a model is more complex.[12] On the one hand, it contains all the models of molecules, and since it also contains all the chemical relations of substances, it must encompass all the different kinds of models of molecules that indicate their various chemically relevant properties. It is clear from Schummer's (1998) report that chemical space is highly idealized and abstract. The idealizations are the very same molecular models: as Schummer points out, substances in real-world quantities do not hold such molecular structures, instead, they are wildly fluctuating. For instance, water, instead of consisting entirely of H_2O molecules, also contains H_3O^+ and OH^- ions (Schummer 1998, 137). But as nodes in that space they must be presented as idealized forms of simple substances—as definitively structured molecules. All the real-world circumstances, such as contaminants (no real sample is fully pure) and other factors that bring "noise" into the stuff, are abstracted away. Not to mention that a purportedly, or practically, pure substance (i.e., pure as defined in the confines of a chemistry laboratory) is a rare occurrence in the real world—most lumps of stuff are kinds of mixts by design or naturally.

The relation of chemical space to actual, spatiotemporal lumps of stuff is at least threefold, which is relevant to mention here:

1) Actual lumps of stuff, first, may refer to a single, ideal entity in the chemical space. That entity may have many embodiments in the real world. Second, actual lumps of stuff usually refer to several entities in the chemical space, as they are mixtures or contain impurities. In those respects, chemical space resembles any scientific model that Giere may think of.

2) The plurality of embodiments of a single element of the chemical space may be due to its various uses; for instance, talc used to lubricate dance floors, talc used as a filler ingredient in pills. From that, one might define another kind of ideal space, a chemical-technological space, containing elements {substance x_n plus its use $un_i_{()}$}, where n, i are natural numbers, or something of the sort (hence including contiguous scientific and engineering fields).[13]

3) Its relation to the chemical practice (of production) is twofold, as mentioned above: the space (its elements) is a result of actual experimental practice, and it guides (further) production. Even if any Gierean model does the same implicitly, chemical space does those things explicitly.

Furthermore, these relations between substances, along with operations through which they can be obtained, and those that can be performed on them to achieve certain aims, are elements of the chemical space, too. This means that the real-world system represented by the chemical space as a model is the chemical core of the *practice* of chemical science, not merely of its results.[14]

Although Giere emphasizes the practical use of models, models are still the center of his approach, and this means that theory is central since models constitute theories, and models resemble but are not equal to reality. Mets (2018) argues for reversing Giere's schema "model resembles reality" into "reality is made to resemble models" for engineering and experimentation.

Her conception of models in these cases resembles Vihalemm's account of laws (Vihalemm 2016b and personal communication) which prescribes how experiments or devices should act upon the world in order for the law to be valid. Thus, for Mets and Vihalemm, reality is created in experimentation and engineering so that it resembles models or laws. I argue that this is applicable to chemical space as well: as a theoretical body consisting of a set of models, chemical space can be used to plan chemical operations and predict their outcomes, leading to the creation of certain substances and relations in the world. Therefore, there is no contradiction between the creative nature of chemistry and the model-based constructive realism of science. Nonetheless, Giere rejects the notion that the world could represent a model (Ronald Giere, personal communication). This idea of the "world representing a model" can serve as an experiment that is built upon the example of a model of an experiment (an element of Giere's hierarchy of models; e.g., Giere 2010). In this case, this "world representing a model" includes the chemical operations carried out in the way that the chemical space—the model—prescribes.

In my adaptation of Giere's schema (figure 10.3), then, the relation between agent and world is "creating {multitude of 'worlds'}," where what for brevity is called "worlds" are the ("pure") substances, the activities of their creation, and the various relations with other substances and the world that they participate in. And as this world of substances is constantly expanding, their world of relations expands too. While Giere does not deny multiple "worlds," his original schema would apply to each substance separately (as mentioned above, e.g., the Lewis structure and others), and he does not admit the creation of "worlds" which here is the object of chemical space as model.

The purpose of the agent should now also be expanded to include the aim of creating new substances in chemistry. Schummer (1997b, 129–30, 131–32)

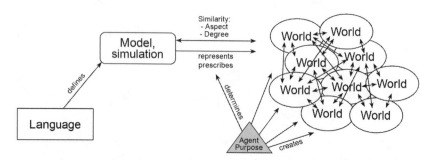

Figure 10.3 Adaptation of Giere's Schema of Model to Chemistry as a Science Creative of Plurality and Focused on Practice. "World" stands for the modeled part of the world (the phenomenon, system, object, attribute). "Language" stands for the terminology and mathematics included in a theory that defines the model. *Source*: Image created by Ave Mets; computer graphics: Margus Evert.

enumerates and specifies the ones mentioned in academic articles, published in the journal *Angewandte Chemie*: synthesis holds the most importance, with technological application not far behind, while theory, classification, and structural typology carry less weight. In adjacent fields, such as materials science, pharmacology, practical application, of course, takes precedence. The latter, in fact, aligns with my argument, which is partly rooted in the plurality of relations in the world that chemical substances (can) engage in, for those are the fields that transform the potential relations of new substances with the world into reality. My argument regarding the creative aspect of chemistry affirms Hasok Chang's[15] thesis that "the *production* of new phenomena is also a most important aim of science, apart from how well we describe and understand them" (Chang 2017, 177).

Giere's perspectival pluralism (2006a, 2006b) is motivated by the fact that science deploys various material research methods and tools to study a single aspect or part of the world, rendering various representations of it and generating a plurality of models. In chemistry, substances themselves are part of the apparatus—they are used in synthesis and the study of other (new) substances. Different apparatuses and varying reaction conditions elucidate different aspects of a substance, modeled, for example, after the different molecular models mentioned above.[16,17] If we allow this pluralism of representations to cover the practices and operations included in chemical space (perhaps in contrast to Giere, who might only include representations of objects and natural phenomena), can we broaden his perspectival pluralism to encompass the creative plurality of chemistry? Giere's targeted pluralism in science is a result of tools employed, and the plurality of those tools depends on the possible aspects and relations of things that the agent deems relevant to their aims. It is primarily a qualitative plurality, as different tools uncover different qualities, and a relatively small amount of research objects is therefore sufficient for generating a qualitative plurality of models. The expansion of chemical space, on the other hand, is primarily quantitative, as it increases both the set of research objects and the tools used to study them. It does bring along qualitative growth, though, possibly in a contiguous or dependent field, when a substance appears clearly novel in some sense.[18] Yet, aside from qualitative novelty (a chemically novel substance), also the sheer quantity (the growth of the number of elements in a type of substances, e.g., metal-organic compounds, as currently predominant [Llanos et al. 2019; Restrepo 2022]) significantly contributes to qualitative growth, since it creates more possibilities for diverse interactions between substances and a greater spectrum of characteristics thereof, which may lead to the emergence of previously unknown properties, even if of a lesser degree of novelty. Hence, while Giere's perspectival pluralism is applicable to chemistry as a creative science, it does not necessarily capture its essence.

CONCLUSION

I have discussed chemistry as a creative science, in which creating a multitude of substances as both research objects and tools is essential, using the framework of constructive realism. I have demonstrated that, first, Vihalemm's concept of φ-science does not neatly classify this aspect of chemistry, as it rather falls somewhere between φ-science and non-φ-science, thus revealing a greater degree of pluralism than Vihalemm contended. Second, while his practical realism proves essential, it is insufficiently specific to understand this plurality and its endorsed pluralism.

Furthermore, I have argued in this section that Giere's model-based account of science, which I use as this missing specification, which also seeps into his perspectival pluralism, must be adapted to take into account chemistry's use of chemical space as an all-encompassing model to understand chemistry as a creative science. The chemical space represents not only the scientific results, the regularities of nature as the object of inquiry, but also chemical operations, and explicitly prescribes what is to be done with the chemical world in order to achieve certain chemical objectives. Although a certain creative relation between the agent and the world applies in both chemistry and physics, given that experimental setups are man-made in both fields, the modification of Giere's account of models is particularly justified when it comes to understanding the essence of chemistry as a quantitatively creative science.

As a result, I have tested, against the essentially ontologically pluralist core of chemistry, approaches to science that either are founded on characteristics of chemistry or claim to capture the character of any scientific field, and have shown that they are insufficient. Chemistry serves as an example justifying pluralism in science, in line with the insights expressed by Kellert et al. (2006, xv): "by the complexity of nature, . . . and the diversity of investigative, representational, and technological goals" (I would include scientific technologies, such as analysis and synthesis). While constructive realist frameworks prove insufficient to fully account for this, philosophy of technology, on its part, fails to capture the scientific epistemic goals. I conclude that further development of practical and constructive realism is necessary for a more comprehensive account of chemistry.

NOTES

1. Many thanks go to Klaus Ruthenberg for his valuable contribution to the discussions that have greatly informed this chapter, presented at the Congress of Logic, Methodology, and Philosophy of Science and Technology 2019 (August 5–10, 2019,

Prague). I extend my gratitude to the participants for the discussions, which have helped to refine the presentation of the topic. The work has been supported by the University of Tartu grants PHVFI16941 and PHVFI20930, the Estonian Research Council grants nos. IUT20-5 and PRG462, and the European Regional Development Fund (Centre of Excellence in Estonian Studies).

2. Lombardi and Labarca (2005, 126) give evidence for and denounce the view that chemistry reduces to physics.

3. Nowadays there is an increasing reliance on computer simulations in making predictions, which indicates that chemistry leans toward becoming even more φ-scientific, as these computer simulations demand precise numerical, digitized models, or algorithms.

4. See, for example, authors such as van Brakel (2012), Chang (e.g., Chang 2012), Lombardi (e.g., Lombardi and Labarca 2005), Ruthenberg (2016; Ruthenberg and Martinez González 2017; Ruthenberg and Mets 2020), Schummer (1997a, 1997b, 2015).

5. Available at https://www.cas.org/support/documentation/chemical-substances. See Hoffmann and Gastreich (2019) for examples of registries specialized on pharmaceutical substances.

6. Vihalemm (2016a, 93) conflates non-φ-scientificity with technologicalness, but this does not make sense, since natural sciences are not technological per se—they do not produce things, production is not their aim. His likening them to each other stems from their complexity: both engage more directly with the real material world (as opposed to engaging with it indirectly and directly with theoretical models instead), and hence have to take its idiosyncrasies and details into account (personal communication). Here, I emphasize the most important features of technology that differentiate it from science: the stance toward material reality as a resource for technical manipulation to serve human needs (for whatever purpose that is relevant in a particular context).

7. See also Schummer (1998).

8. See also Chang (2017).

9. See note 6.

10. See Mets (2018).

11. I thank Alar Sünter for this example.

12. See also Llanos et al. (2019) and Restrepo (2022) on the chemical space.

13. Trivially, classes of any classification system have many instantiations in the world, such as many specimens belonging to a particular species (regardless of the definition of "species"). In terms of biological classes, all specimens share the basic functions of a specimen (breathing, nutrition, procreation, etc.), while the function that talc serves as a filler is different from its function as a lubricant.

14. See van Brakel (2012, 198).

15. Chang (2022) advocates for the same view, but this chapter could not benefit from it. Thanks to an anonymous referee for bringing this to my attention.

16. See also Llored (2015).

17. For instance, helium as a noble gas was long thought to be chemically entirely inert, but under high pressure, it can form compounds, for example, with sodium.

18. I acknowledge that this judgment depends on what we consider a qualitatively novel aspect or relation of a chemical substance. For instance, the first antidepressant was novel because there were no chemical substances capable of influencing the brain in such a manner before. But another antidepressant, that also impacts brain chemistry, but in a markedly different manner—is it then a novel relation in the world? This discrimination may be arbitrary, or it could depend on the purpose of discriminating levels of novelty. Both of these substances may have been known and accepted in the chemical space already (as, for instance, lithium was), thus contributing to its quantity, but exploring the gamut of their properties and relations with the myriad denizens of this world, including their effects on the neural system, is a gradually unfolding process.

REFERENCES

Bachelard, Gaston. 1968. *The Philosophy of No: A Philosophy of the New Scientific Mind.* New York: The Orion Press.

Bensaude-Vincent, Bernadette. 2012. "Gaston Bachelard (1884–1962)." In *Philosophy of Chemistry: Handbook of the Philosophy of Science*, edited by Andrea I. Woody, Robin Findlay Hendry, and Paul Needham, 142–50. Amsterdam: Elsevier. https://doi.org/10.1016/B978-0-444-51675-6.50013-X.

Bensaude-Vincent, Bernadette, and Jonathan Simon. 2008. *Chemistry—The Impure Science.* London: Imperial College Press. https://doi.org/10.1142/p569.

Berthelot, Marcellin. 1876. *La synthèse chimique.* Paris: Alcan.

Chang, Hasok. 2012. *Is Water H_2O? Evidence, Pluralism and Realism.* Boston Studies in the Philosophy of Science. Dordrecht: Springer. https://doi.org/10.1007/978-94-007-3932-1.

Chang, Hasok. 2017. "What History Tells Us About the Distinct Nature of Chemistry." *Ambix* 64 (4): 360–74. https://doi.org/10.1080/00026980.2017.1412135.

Chang, Hasok. 2022. *Realism for Realistic People: A New Pragmatist Philosophy of Science.* Cambridge: Cambridge University Press. https://doi.org/10.1017/9781108635738.

Giere, Ronald N. 1988. *Explaining Science: A Cognitive Approach.* London and Chicago, IL: University of Chicago Press. https://doi.org/10.7208/chicago/9780226292038.001.0001.

Giere, Ronald N. 2004. "How Models Are Used to Represent Reality." *Philosophy of Science* 71 (5): 742–52. https://doi.org/10.1086/425063.

Giere, Ronald N. 2006a. "Perspectival Pluralism." In *Minnesota Studies in the Philosophy of Science*, vol. XIX, edited by Stephen H. Kellert, Helen E. Longino, and C. Kenneth Waters, 26–41. Minneapolis, MN: University of Minnesota Press.

Giere, Ronald N. 2006b. *Scientific Perspectivism.* London and Chicago, IL: University of Chicago Press. https://doi.org/10.7208/chicago/9780226292144.001.0001.

Giere, Ronald N. 2010. "An Agent-Based Conception of Models and Scientific Representation." *Synthese* 172: 269–81. https://doi.org/10.1007/s11229-009-9506-z.

Hoffmann, Torsten, and Marcus Gastreich. 2019. "The Next Level in Chemical Space Navigation: Going Far Beyond Enumerable Compound Libraries." *Drug Discovery Today* 24 (5): 1148–56. https://doi.org/10.1016/j.drudis.2019.02.013.

Holum, John R. 1996. *Organic and Biological Chemistry*. New York, Toronto: John Wiley and Sons.

Kellert, Stephen H., Helen E. Longino, and C. Kenneth Waters. 2006. "Introduction: The Pluralist Stance." In *Scientific Pluralism*, edited by Stephen H. Kellert, Helen E. Longino, and C. Kenneth Waters, vii–xxix. Minneapolis, MN: University of Minnesota Press.

Llanos, Eugenio J., Wilmer Leal, Duc. H. Luu, Jürgen Jost, Peter F. Stadler, and Guillermo Restrepo. 2019. "Exploration of the Chemical Space and Its Three Historical Regimes." *Proceedings of the National Academy of Sciences* 116 (26): 12816660–65. https://doi.org/10.1073/pnas.1 039116.

Llored, Jean-Pierre N. 2015. "Investigating the Meaning of Ceteris Paribus Clause in Chemistry." In *Philosophy of Chemistry: Growth of a New Discipline*. Boston Studies in the Philosophy and History of Science 306, edited by Eric Scerri and Lee McIntyre, 219–31. Dordrecht, Heidelberg, New York, and London: Springer. https://doi.org/10.1007/978-94-017-9364-3_14.

Lombardi, Olimpia, and Martín Labarca. 2005. "The Ontological Autonomy of the Chemical World." *Foundations of Chemistry* 7: 125–48. https://doi.org/10.1007/s10698-004-0980-6.

Mets, Ave. 2018. "Normativity of Scientific Laws (I): Two Kinds of Normativity." *Problemos* 93: 60–69. https://doi.org/10.15388/Problemos.2018.93.11751.

Restrepo, Guillermo. 2022. "Chemical Space: Limits, Evolution and Modelling of an Object Bigger Than Our Universal Library." *Digital Discovery* 1: 568–85. https://doi.org/10.1039/D2DD00030J.

Ruthenberg, Klaus. 2016. "Matter and Stuff—Two Sides of the Same Medal?" In *Understanding Matter Vol. 2. Contemporary Lines*, edited by Andrea Le Moli and Angelo Cicatello, 153–68. Palermo: New Digital Press.

Ruthenberg, Klaus, and Jaap van Brakel. 2008. *Stuff: The Nature of Chemical Substances*. Würzburg: Königshausen & Neumann.

Ruthenberg, Klaus, and Juan C. Martinez González. 2017. "Electronegativity and Its Multiple Faces: Persistence and Measurement." *Foundations of Chemistry* 19: 61–75. https://doi.org/10.1007/s10698-017-9278-3.

Ruthenberg, Klaus, and Ave Mets. 2020. "Chemistry Is Pluralistic." *Foundations of Chemistry* 22: 403–19. https://doi.org/10.1007/s10698-020-09378-0.

Schummer, Joachim. 1996. *Realismus und Chemie. Philosophische Untersuchungen der Wissenschaft von den Stoffen*. Epistemata. Würzburger wissenschaftliche Schriften. Reihe Philosophie, Band 178. Würzburg: Königshausen & Neumann.

Schummer, Joachim. 1997a. "Scientometric Studies on Chemistry I. The Exponential Growth of Chemical Substances, 1800–1995." *Scientometrics* 39 (1): 107–23. https://doi.org/10.1007/BF02457433.

Schummer, Joachim. 1997b. "Scientometric Studies on Chemistry II. Aims and Methods of Producing New Chemical Substances." *Scientometrics* 39 (1): 125–40. https://doi.org/10.1007/BF02457434.

Schummer, Joachim. 1998. "The Chemical Core of Chemistry I: A Conceptual Approach." *Hyle* 4 (2): 129–62.

Schummer, Joachim. 2015. "The Methodological Pluralism of Chemistry and Its Philosophical Implications." In *Philosophy of Chemistry—Growth of a New Discipline*, edited by Eric Scerri and Lee McIntyre, 57–72. Dordrecht: Springer. https://doi.org/10.1007/978-94-017-9364-3_5.

Simon, Jonathan. 2012. "Chemistry and Pharmacy: A Philosophical Inquiry into an Evolving Relationship." In *Philosophy of Chemistry: Handbook of the Philosophy of Science*, edited by Andrea I. Woody, Robin Findlay Hendry, and Paul Needham, 519–30. Amsterdam: Elsevier.

Van Brakel, Jaap. 2000. *Philosophy of Chemistry: Between the Manifest and the Scientific Image*. Leuven: Leuven University Press.

Van Brakel, Jaap. 2012. "Substances: The Ontology of Chemistry." In *Philosophy of Chemistry: Handbook of the Philosophy of Science*, edited by Andrea I. Woody, Robin Findlay Hendry, and Paul Needham, 191–29. Amsterdam: Elsevier. https://doi.org/10.1016/B978-0-444-51675-6.50018-9.

Vihalemm, Rein. 2011a. "A Monistic or a Pluralistic View of Science: Why Bother?" In *An den Grenzen der Wissenschaft*. Die "Annalen der Naturphilosophie" und das natur- und kulturphilosophische Programm ihrer Herausgeber Wilhelm Ostwald und Rudolf Goldscheid, edited by Pirmin Stekeler-Weithofer, Heiner Kaden, and Nikolaos Psarros, 79–93. Stuttgart, Leipzig: Sächsische Akademie der Wissenschaften zu Leipzig.

Vihalemm, Rein. 2011b. "Towards a Practical Realist Philosophy of Science." *Baltic Journal of European Studies* 1 (1): 46–60.

Vihalemm, Rein. 2016a. "Chemistry and the Problem of Pluralism in Science: An Analysis Concerning Philosophical and Scientific Disagreements." *Foundations of Chemistry* 18: 91–102. https://doi.org/10.1007/s10698-015-9241-0.

Vihalemm, Rein. 2016b. "Science, φ-Science, and the Dual Character of Chemistry." In *Essays in Philosophy of Chemistry*, edited by Eric Scerri and Grant Fisher, 352–79. New York: Oxford University Press. https://doi.org/10.1093/oso/9780190494599.003.0024.

Chapter 11

Molecular Biology and Genetics

On the Evolution of the Nature of Modern Biology and Its Relation to Chemistry and Physics

Apostolos K. Gerontas

The past decades have witnessed explosive growth in areas of biology and its medical applications, primarily through the introduction of improved instrumentation and techniques originating in chemistry and physics, alongside concepts and research approaches.[1] The process of chemicalization and physicalization of the biological domains did not just lead to the formation of molecular biology and genetics but altered the whole plateau of biology by transforming the available vocabulary and the nature of its discourses. This chapter examines the historically critical points of this transformation and the relevant intra-disciplinary discourses, contributing to the discussion on the new nature of biology. Molecular biology and genetics are examined philosophically through the lens of Rein Vihalemm's φ-science: Does biology have a dual nature, like the one Vihalemm described for chemistry?

Chemistry and biology have had a relationship with one another that stretches back to the eighteenth century. The problems being addressed by the two disciplines were intimately intertwined, already in their earliest forms. Bridged all by medicine, the studies of chemistry, natural history, and physiology were often pursued by the same individuals, in the same rooms, and serving similar agendas. The effect that the studies of life had on the development of modern chemistry is often underrepresented in both historical and philosophical literature. New Chemistry began with an act of emancipation, not from the alchemical practices and theories, but from its medical bonds. Chemists of the late eighteenth century, especially in the then pioneering France, wished to enter the high salons of the *philosophes* of the Enlightenment. To this purpose, a type of higher "chemical philosophy" (or "philosophical chemistry") was necessary, and the bonds of their discipline to more practical considerations—such as pharmacy—needed to be severed (Simon 1998).

The work of Antoine-Laurent de Lavoisier (1743–1794) can be viewed as the culmination of this path. This dimension of emancipation from medicine and physiology is almost always left out in the histories of the so-called *chemical revolution*, and it is also not touched upon by Rein Vihalemm in his known work, *A Story of a Science* (Vihalemm 2019). This omission—done by historians and philosophers alike—is obscuring the relations of chemistry with natural-historical practices, physiology, and medicine and might be causing misunderstanding concerning the exact nature of chemistry as the "central science."[2] Still, it could be said that the ideal construct of "philosophical chemistry" (or the other way around, "chemical philosophy") does roughly correspond to future notions of an ideal "exact science" part in chemistry—or Rein Vihalemm's *φ-science* nature of chemistry.

Chemistry, of course, was never really separated from physiology, despite the intentions of the chemical philosophers of the late eighteenth century. The relationship of the two disciplines continued into the twentieth century and became more intimate. During this century, chemistry underwent its second revolutionary period, restructured by the introduction of physical and physical-chemical analytical methods, more often than not in their mechanized-automated incarnations (chromatographs, spectrometers of different types, elemental analyzers of different types, etc.). With a short delay, biology was also rearranged with a newfound focus on the molecular level. During the past decades, molecular biology laboratories became virtually indistinguishable from chemical laboratories in terms of equipment and research routines.

On the introduction of the physical and physical-chemical methods in chemistry, as well as the process and the effects of their automatization (the so-called *instrumentation revolution*), there is a continuously growing literature.[3] Elements of the transformation of chemistry during the twentieth century have themselves been instigated by the needs of biological research: chromatography, for example, was first introduced to be used in research on chlorophyll and was reintroduced to enable research on carotenoids. Subsequently, also high-performance liquid chromatography was created to serve the needs of biochemists (Gerontas 2013, 2020). There is also a growing literature on the history of molecular biology and genetics,[4] but the literature becomes scarcer when it comes to the evaluation of the changes in biology due to the increased chemicalization and automatization. It is relatively clear that the automation introduced in chemistry increased its potential uses in biological studies (simply by offering unprecedented speed and, later, computing capacity). Whether these changes have indeed made biology "more of a science, and less of a descriptive discipline," as a chemist claimed to the author almost a decade ago, however, is the subject of the current chapter.[5]

THE GENE: FROM ITS CONCEPTION
TO ITS MOLECULARIZATION

Enter the twentieth century, the studies of life had attained crucial tools of explanation and a vocabulary to fulfill their descriptive role concerning the phenomena of inheritance. Decades of evolutionary thought, combined with the laws of inheritance of Gregor Mendel (1822–1884), have contributed to the exactness and robustness of scholarly descriptions by taking these descriptions one level lower than the observed biological characteristics and offering them transgenerational explanatory power.

Mendel's results, rediscovered and reevaluated decades after their time, ushered in the era of classical genetics. Conducted between 1856 and 1863, presented in 1865, and published in 1866, these experiments were at their time considered to be on hybridization not inheritance—Mendel himself considered his results to apply only to a narrow number of traits and a few species only (Klein and Klein 2013; Fr. Richter 2015). After the rediscovery of his work in 1909, the Danish botanist and pharmacist Wilhelm Johannsen (1857–1927) coined the term *gene* to describe the units of heredity that were proposed by Mendel. The word *genetics* had already been coined, a few years back, by the English Mendelian William Bateson (1861–1926), but his term needed the existence of the word gene to catch up. Wilhelm Johannsen also coined the terms *phenotype* and *genotype* (Johannsen 1911). From that point on, the term gene took a life of its own and would have to be reinvented several times (and is probably, once again, in need of a redefinition, see Portin and Wilkins 2017).

Initially, the concept of gene was intended to be a mere abstraction. Mendel had indeed considered the possibility of the existence of minuscule factors that determined inheritance ("cell elements," or *Zellelemente*; Portin and Wilkins 2017), but Johannsen thought of the gene as a Mendelian hereditary factor and avoided speculating on its physical nature and structure. A couple of decades later, however, some genes had indeed been localized and could be treated as points on the physically existing and visible chromosomes. Two decades further on, the genes have acquired the dimension of length—they have been shown to have an internal structure and to be dissectible—and, by the 1960s, they had acquired a three-dimensional structure as chemical entities, that is, chemical molecules (Portin and Wilkins 2017).

During the first three decades of the twentieth century, biologists established a muster of genetic experimentation—which included experimentation on the heroic species *Drosophila melanogaster*, which should be honored by statues in front of biology and medicine departments—and developed a quasi-Mendelian model. Mathematicians, on the other hand, introduced statistics in the studies of heredity, making population genetics a possibility even for humans. The developments opened the way for phenotypical studies and studies on general populations. Alongside these developments, the step-by-step deciphering of biochemical systems led researchers to something that was first an ideal conception of the relationship between biological systems and chemistry, until it eventually became a field—*molecular biology*.

Already, Darwin had imagined corpuscles that were responsible for the transmission of hereditary characteristics. The American geneticist Hermann Joseph Muller (1890–1967) took an important step further in the 1920s. Genes were ultramicroscopic particles that had the peculiar quality of self-propagation, he thought, and they were the vehicles of evolution, the *very basis of life* (Muller 1926), and that: "when the structure of the gene becomes changed, through some 'chance variation,' the catalytic property of the gene may become correspondingly changed, in such a way as to leave it still auto-catalytic" (Muller 1922; also reported in Falk [2010]).

Muller's view, which logically implied the reducibility of biology to chemistry, was programmatic and soon popular. Biochemists, in particular, were to prove enthusiastic about the view: the concept of molecular biology was thus more solidly defined as "the biochemistry related to DNA and its expression into proteins" (Rheinberger 1997; Kellenberger 2004), allowing for the reshuffling of the disciplinary cards and the accommodation of a multitude of biochemists in this research agenda. The reductionist approach was, after all, also epistemologically promising.

The biology-to-chemistry view had great success early on. Molecular methods carved new paths for the understanding of evolutionary processes. New concepts such as the molecular evolutionary clock, as well as the discovery of repetitive sequences in eukaryotic genomes, demonstrated that evolution might be occurring differently at the organismic and molecular levels, sparking debates between defendants of the molecular approaches and supporters of organismic views (Suárez-Díaz 2016).

In the 1960s, these early successes made molecular biology visible to governments, funding bodies, and eventually the public (Rheinberger 2012). With this visibility came not only funding but also a disciplinary identity and supporting organizations. Progressively, chemistry and molecular studies in biology came to deploy similar conceptual tools and vocabularies—partially due to the reductionist agenda of many researchers, and partially due to the general interdisciplinarity imposed by governmental and industrial funding bodies (a process that had similarly happened in some areas of chemistry, roughly a generation before). And, as it happened with chemistry earlier, the issue of the reducibility or not generated discussions, both between practitioners and epistemologists.

NATURAL HISTORY AND SCIENCE: Φ-SCIENCE FROM CHEMISTRY TO BIOLOGY

In earlier discussions about the relationship between chemistry and physics, scientists and philosophers held quite diverse views that represented different areas of emergence phenomena and differing levels of reducibility. Scerri (2012), for example, examined this relation on the level of elements in the periodic table; Hendry (2010) examined this relation on the level of molecular structure and concluded that "emergence is obscure and of doubtful coherence," and that emergence is "metaphysically impossible"; and Hettema (2012, 2017) saw a "unity between chemistry and physics" through the ontology of quantum chemistry as a zone "in-between."

The discussion remains complex: chemistry and physics encompass various phenomena, levels of explanation, competing theories and views, as well as methods, research aims, and programs. Thus, while in one case the path from physics to chemistry might be adequately describable, in the other it is not. Attempts to describe all the chemical bonds as a special case of an emergent property seem to be mathematically promising,[6] but it still seems that the observable chemical phenomena are only partially reducible to the quantum level. There are still chemical properties that cannot be sufficiently understood in terms of reduction or emergence. Among them, the molecular structure is among the most fundamental.[7]

Rein Vihalemm tackled the issue of the demarcation of chemistry and physics through the lens of the philosophy of practical realism. Irrespective of the ramifications of quantum physics theory and the development of the quantum chemistry bridge, Vihalemm thought that chemistry still differed from physics on the level of practical aims, methods, and activity. As far as he saw it, chemistry was motivated by considerations that were significantly more local, and its generalized theories and systems' descriptions were intended as solutions to given problems and not as general theories of all. On that, Vihalemm was voicing views that were privately expressed by a multitude of chemists of the said period: "Whether a reaction system is described by quantum mechanics or not," said a chemist to the author once, "it is irrelevant, as long as the system functions as intended."[8] For the everyday life of a chemist in the laboratory, predicting the outcomes of a reaction is significantly more important than the underlying reality of this reaction, and this practical focus creates a research program radically different from any one generated in the disciplinary environment of physics. At least externally, a demarcation of the disciplines remains possible—and is practically wished too.

Vihalemm identified in chemistry and its practices two coexisting "natures" (a historian would, probably more appropriately, write "two historical traditions"). A part of chemistry, Vihalemm thought, is indeed a "proper science in the sense that physics is" a *φ-science*. Another part of chemistry, however, is akin to the natural history of the past—and biology before its molecularization, we could add. This part of chemistry is gathering descriptions, cataloging, and classifying entities. In Vihalemm's words, φ-science has a specific aim and function:

> The aim of φ-science is to examine reality from the viewpoint of laws of nature. Examining reality from such a viewpoint presupposes the construction of models as experimentally substantiated idealizations. (Vihalemm 2007, 231)

Vihalemm stressed that the concept of φ-science is an idealization and emphasized the use of the concept specifically for the demarcation of chemistry and physics. As such, however, this idealized concept is practically identical to a more general and standard *Vorbild* of an exact science—similar to those proposed by Hempel, Popper, and Nagel—and is often known as the "standard conception" of scientific theories.[9] While there are alternatives, the author found this view of scientific structure most useful, while thinking about the developments in biology of the past decades.

The reader might be tempted to remember what was mentioned earlier in this chapter: the theoretical program of Lavoisier and his school included the severing of the bonds of chemistry to medicine (and its natural-historical roots) and the building up of philosophical chemistry that would take its place in the Enlightenment salons. Thus, already at the beginning of modern chemistry, some practitioners saw chemistry as dual-natured and strove to strengthen the one nature that was perceived as "purer," "nobler," and more scientific (or, using the terminology of that era, "more philosophical").

A particular characteristic of φ-science, as framed by Vihalemm, is that it corresponds to a very specific way of seeing the world: what cannot be described in its idealized laws is invisible and unphrasable in φ-science terms (Näpinen 2015, 109). On the other hand, a field of natural history describes and catalogs entities on a more local level, and without the need to solidly explain and justify their existence. The φ-science concept would then be of value in the case of the discussion around the molecularization of biology over the past decades. The studies of life (including medicine) had been one of the last fortresses of natural-historical views, practices, and research programs. Botany, zoology, microbiology, and the related disciplines of pharmacology and early immunology had had a long history of collecting and classifying before the advent of molecular views. It could even be claimed that their actual epistemic objects were primarily the catalogs of observed properties that they were generating and, only secondarily, actual physical objects, such as animals, diseases, patients, or collections. During the twentieth century, however, and with the introduction of molecular practices and vocabulary, there was a fundamental shift in the focus of the life sciences. On the molecular level, generalized laws could be now pronounced that were valid across species, genera, domains, or (why not) even life-supporting planets. That led to the strengthening of a part of biology that was weak until then and altered the aspirations of its practitioners.

BIOLOGY, "NEW" AND "OLD"

Some of the early successes of molecular biology were so promising as to captivate the public mind and generate pop phenomena—science-fiction stories, films, comic book storylines, and franchises. The word "mutant" (probably existing since 1900) entered science fiction in 1954, and the X-Men comic series (where genetics was interestingly intertwined with the atomic age and a particularly American racial issues metaphor) was launched by Marvel Comics in 1963. Philip K. Dick's "androids" from the 1968 novel *Do Androids Dream of Electric Sheep?*—initially conceived as robots made out of organic matter only looking indistinguishable from humans—were revamped as living "replicants" in the *Blade Runner* film adaptation of 1982. By 1997, the film industry had produced dystopic versions of the future, where only genetically "improved" humans would enjoy full rights and access to prestigious professions—the world presented by the film *Gattaca*. "If the 20th century was the century of physics, the 21st century will be the century of biology," was even boldly proclaimed (Venter and Cohen 2004).

Venter and Cohen did not mean all of biology, of course, and they certainly did not mean this part of biology that the world knew before the 1940s. In their paper, the two geneticists were clear: they were talking about the "new biology of genome research" (there was no need to emphasize the word *new* here), which offers "a *complete description* of life at the most fundamental level of the genetic code" (here, however, the words *complete description* were emphasized by the author). Not only this, the two wrote, but "also the precoded information," or "chemical spelling" which is responsible for the on or off positions of genes will be open for us. Soon, we shall be able to know genetic predispositions for persons, and even manipulate genes "to produce blue eyes or dark skin."

Leaving aside the unlucky choice of words in the last sentence, two things were made clear by Venter and Cohen's bold pronouncements: (a) the brave and bold future belonged to the *new* biology, not the old; and (b) the ideas reaching the public through science fiction stories and films were partially shared by leading members of the (molecular) genetics community. The "new" biology was, in fact, chemistry: analytic at step one, and then synthetic—it had not just absorbed methods and vocabulary from chemistry, it had absorbed a research program and aims as well.

That things would soon develop this way was clear among members of the biological community quite early on—and the potential effects were also clear. In 1964, already, the famous geneticist and evolutionary biologist Theodosius Grigorievich Dobzhansky (1900–1975) held an important speech on exactly the question of the relations between the organismic views of the old, and the then young molecular biology of his time (Dobzhansky 1964). Disturbed by the generally spreading belief that the only biology worth pursuing was the molecular one, Dobzhansky was drawing the line of defense for the old biology:

> Biology is structured rather differently from other natural sciences. . . . A biologist, more than, for example, a physicist or a geologist, is faced with several hierarchically superimposed levels of integration in the objects which he studies. Life presents itself to our view almost always in the form of discrete quanta—individuals. But unlike the atoms of classical physics, individuals are conspicuously divisible, and, unlike the atoms of modern physics, divisible into great numbers and a great variety of component elements, cells. . . . Chromosomes and genes have that extraordinary chemical substance, the DNA, as the key constituent. But the DNA in the chromosome is something more than the DNA in a test tube. A chromosome is an organized body, and its organization is as essential as is its composition. (Dobzhansky 1964)

Biology should not be treated as the chemistry of living things, and structure and organization are equally important to chemical composition. The classifications of the old biology should also keep their importance:

> The supra-individual forms of integration seem less tangible in a spatio-temporal sense than the infra-individual ones, but just as interesting and significant. Mankind is less clearly perceived by our sense organs than an individual man, but it is nevertheless as meaningful a biological entity as it is a cultural entity. The sexual mode of reproduction connects individuals into reproductive communities, Mendelian populations. Mendelian populations are united by reproductive bonds into inclusive reproductive systems—biological species. An isolated individual, especially an individual of a sexual species, is at least as clearly an anomaly as a cell isolated from a multicellular body. (Dobzhansky 1964)

Dobzhansky insisted, above all, on the preservation of the importance of the organismic level. He thought that a trace of positivism (of the Auguste Comte variety) was hidden behind the reductionist project—an unspoken belief that chemistry and, above all, physics are somehow more advanced sciences than biology. In these terms, then, the reductionist project was aiming to effectively make biology an obsolete discipline by reducing it first to chemical and, eventually, to physical explanations.

And yet, Dobzhansky's aim was not to say that biological phenomena cannot ever be completely reduced to chemistry and physics. Quoting a relevant text by Ernest Nagel (1901–1985), he reasoned that biology *will* be eventually reducible to chemistry and physics—but *not just yet*:

> The progress of biology would not be furthered by frenetic efforts to reduce organismic biology to chemistry or physics. This is not because there is anything in living things that is inherently irreducible. It is rather because a different research strategy is more expedient. Those who urge an immediate absorption of the organismic into molecular biology neglect the simple but basic fact that life has developed several levels of organization. These are levels of increasing complexity, and they are hierarchically superimposed. The elementary phenomena and regularities on each succeeding level are organized patterns of those on the preceding level. Organismic biology can be said to be a study of patterns of molecular phenomena. Such a definition of organismic biology is correct as far as it goes, but it does not go quite far enough. It is a study not only of the molecular patterns but also of patterns of patterns. (Dobzhansky 1964)

Between 1964 and 2004, and between Dobzhansky and Venter and Cohen, it is unclear whether the appropriate time for this pre-announced reduction has arrived—but the distance between the old biology and its methods and the new molecularized biology has been furthered. Indeed, even population biologists and ethologists were tempted to attribute the most complicated biological phenomena (and, occasionally, even social phenomena) to the genes only (which were mostly selfish, as Dawkins put it in his 1976 book by that name). Simultaneously, the descriptive and classifying subdisciplines of biology were losing in prestige, as they failed to demonstrate their grounding at the molecular level.

HOW CLOSE IS MODERN BIOLOGY TO Φ-SCIENCE ANYWAY?

One of the most striking aspects of the history of biology's molecularization is that this development generated existential discussions over the whole plateau of the life sciences. While, in the case of chemistry, the reducibility or not to physics was (and is) debated, at no point did in the chemical studies arise a view similar to the one that arose in biology while it was being molecularized: that the only biology worth the name of science is the reducible biology.

As Dobzhansky put it in his aforementioned speech:

> The notion has gained some currency that the only worthwhile biology is molecular biology. All else is "bird watching" or "butterfly collecting." Bird watching and butterfly collecting are occupations manifestly unworthy of serious scientists! I have heard a man whose official title happens to be Professor of Zoology declare to an assembly of his colleagues that "a good man cannot teach zoology." A good man can teach, of course, only molecular biology. (Dobzhansky 1964)

These existential issues in biology had deep roots in the development of the discipline—and, indeed, reflected a sense of incompleteness as a science that was generated by the discipline's strong natural-historical nature. In the meanwhile, an idealized and a-historical view of what science is (and what science supposedly does, or can do) had spread among the intellectual elites and the pop culture, affecting the views that biologists held of their discipline and reinforcing their sense of incompleteness. The surprising—panegyric and, simultaneously, self-canceling—reaction of the biological communities to the emergence of molecular biology is then an important historical datum. And an epistemological one.

The mentioned internal sense of incompleteness of biology had, of course, roots in the very formation of biology out of natural history in the eighteenth century. Under the pressures of the successes of astronomy, and aided by mechanistic-atomistic views, natural historians and physicians alike tended to view life on a mechanistic basis. While the mechanistic views seemed to be adding potency and confidence to medicine, they were not fully satisfying. Living systems, after all, possessed properties that were not observable to nonliving ones: they could heal, for example; they could regenerate whole limbs occasionally; and they could procreate without the help of any (at least visible) engineer.

This train of thought eventually generated *vitalism* as a movement: the idea that in living matter there is a semimystical *vis vitalis* that fundamentally separates it from the nonliving one. Vitalism offered significant services in the studies of life—among them, also the coining of the term "biology" (for histories of the controversy between mechanism and vitalism, see De Klerk 1979; Porter 2011). Vitalistic and, later on, systemic and holistic approaches were in any case promoted, in the face of biology's inability to generate purely mechanistic theories concerning the nature of living matter.

Of equal importance was, of course, the external pressure on the discipline. Dealing with the extremely special and distant case of the cosmos, astronomy was able to generate views that seemed particularly consistent, explanations that seemed solid, and predictions that, indeed, were irrefutable. All of these had become, during the Enlightenment, the measure of exactness and were pressuring all the "less exact" disciplines (not only the natural sciences, but also the humanities, including the studies of societies, economy, and ethics). The theories of evolution generated between Lamarck and Darwin were celebrated as concise explanations of the past of life, but they could not relieve the aforementioned external pressure—especially when they were viewed side by side with the advances of chemistry and physics of the same era. Evolution(s) was not exact enough, failed to explain the origin of life, could not be mathematized, and offered no predictions.

This sense of impotence was visible also outside of biology. In the 1930s, physicists, chemists, and mathematicians were drawn to, and troubled by, what was perceived as an arrested development of the studies of life (and the psyche). In some circles, even the financial crisis of 1929 was attributed to this imbalance in development: humans had learned a lot about inanimate objects and nonliving matter but not enough about biology and psychology. These were the areas of study that should be emphasized, alongside physics, chemistry, and mathematics that would be necessary to support the former. Perhaps a new "Science of Man" could be created, some influential circles suggested (Morange 2020, 80–81). Physicists like Bohr and Schrödinger turned their attention to the state of biology and invited their colleagues to do so: that was, after all, the new frontier.[10] And the astonishing early progress that followed seemed to prove their case.

The early advances, however, were soon followed by startling discoveries of inability. Initially seeing causality in biological systems in terms of linear chains of events (the so-called *Descartes' clockwork*), molecular biologists pretty soon had to accommodate multidimensionality (Kellenberger 2004). In complex biological systems, more often than not, linear causation did not work well and had to be replaced by multidimensional meshworks. In such meshworks, linear sequences of causalities intertwine—and "at every point where two or more causalities join, deviations from the chain to other parts of the meshwork are possible" (Kellenberger 2004). On the molecular level, it is highly improbable that a given cause would give a single effect. On the contrary, "side effects or completely changed effects are possible as well" (Kellenberger 2004), a fact that explains the incredible variety of phenotypes produced by similar genotypes. Hence, in the position of classic (φ-science) causality, molecular biology and genetics had to settle with *near-causality*, which only rarely can produce reproducible experimental sequences and must be highly dependent on complicated statistics and powerful computer systems. In biological systems, causal relations can be established only after great amounts of data, representing great numbers of test objects and experimental sequences, are gathered and processed (and, even then, outliers exist).

While the (often imagined or speculated, but always offered) explanations of classical biology in its previous natural-historic form were largely linearly causal, its molecularization weakened its explanatory certainty by drawing the whole discipline toward the complexities of the underlying reality of the biological phenomena. Practitioners dealt with near-causalities on the system level by adopting holistic approaches, taking into account that actions might be generating different feedback in different cases and *a priori* accepting that a knowledge of all the relevant factors might be impossible. Calls for strategies for defining causality in biological systems and achieving desired system-level outcomes exist (Bizzarri et al. 2019), techniques based on multi-omics datasets have been developed,[11] but they collide on limitations probably imposed by the very nature of living systems, even on a quite basic level.

CONCLUSION

Molecular studies of life indubitably offered biology a level of potency that the discipline had never enjoyed before. A new level of analysis had been opened, and, with it, a level of possible intervention was achieved that would have been considered impossible two generations ago. During this period of transformation, biology left behind its previously natural-historical nature, turning its main focus from descriptive-explanatory models to analytic-synthetic explanatory models previously developed in biochemistry. Hence, to the eyes of the external beholder, biology had indeed taken a critical step in the direction of "maturing" to a φ-science—an exact science proper. This change in nature and status has been affirmed by the public, professionals of other disciplines, governments, funding bodies, and (quite importantly) the industry. Although history and philosophical analysis mostly happen in retrospect, it is fair to write that the prediction of Venter and Cohen, mentioned earlier, will have a good chance of becoming true: the twenty-first century *will* most probably be the century of biology.

Historically and philosophically, though, it is worth mentioning that the apparent step toward "exactness" is less pronounced than what is going through to the public opinion. Charles Darwin might famously have lost his sleep over the peacock's tail (namely, he could not understand why evolution turned the male peacock into a moving target), but his evolutionary theory could produce a linear-causal explanation that was at some level logical, true, and generalizable.[12] While on the level of appearances, molecular explanations are significantly more spectacular—and more mathematizable, a fact that plays an important role in the declaration that a science is exact—in reality, they are significantly less straightforward. Linear-causal explanations are almost impossible in molecular biology and genetics, and even theoretically simple systems and problems are more often than not described in vague and uncertain terms.

It could be argued then that molecular biology discovered a *principle of uncertainty* on the system-molecular level, a few levels higher than the one described by Werner Heisenberg in 1927. That, in itself, would be interesting for scientists, historians, and philosophers. It takes an extra level of significance, however, due to the importance that molecular biology and genetics already play in our everyday lives. Near-causality does not only mean that biology is not much of a φ-science. It also means that its readings of causes of observed effects are uncertain and that the effects of its interventions might be even more of the same. Since the study of the living systems is indeed the study of "patterns of patterns," and ecosystems and the planet add up additional patterns, the above conclusion is worth considering.

NOTES

1. Some of the data and thoughts presented in this chapter were generated while the author was a visiting researcher on the history of medicine at the American College of Greece in Athens (2020–2022, a position organized by ARISTEiA and funded by the biomedical companies Algonot and Attica Sciences). Some of the insights were generated during discussions with students of the courses in philosophy of biology and bioethics that the author was teaching at the Coburg University of Applied Sciences in the period 2014–2021. It is impossible to name all of these students, but I thank each and every one of them.

2. The term "central science" was popularized in a textbook in chemistry, first published in 1977, see Brown and LeMay (1977).

3. See Baird (1993), Morris (2002), and Reinhardt (2006).

4. See Kellenberger (2004) and Morange (1998, 2020).

5. Personal communication with a chemist at the Department of Chemistry at the Norwegian University of Science and Technology in Trondheim in 2011, where the author was a doctoral fellow.

6. See, for example, Golden, Ho, and Lubchenko (2017).

7. See Seifert (2022) and Hendry (2017) for arguments of a strong emergence of the molecular structure.

8. Personal communication with an organic chemist, distinct from the individual mentioned in note 5.

9. See, among others, Hempel (1970) and Nagel (1961).

10. See Morange (2020, 67–78).

11. See Kelly et al. (2022).

12. "The sight of a feather in a peacock's tail," he wrote in 1860, "makes me sick."

REFERENCES

Baird, Davis. 1993. "Analytical Chemistry and the 'Big' Scientific Instrumentation Revolution." *Annals of Science* 50 (3): 267–90. https://doi.org/10.1080/00033799300200221.

Bizzarri, Mariano, Douglas E. Brash, James Briscoe, Verônica A. Grieneisen, Claudio D. Stern, and Michael Levin. 2019. "A Call for a Better Understanding of Causation in Cell Biology." *Nature Reviews Molecular Cell Biology* 20 (5): 261–62. https://doi.org/10.1038/s41580-019-0127-1.

Brown, Theodore L., and H. Eugene LeMay. 1977. *Chemistry: The Central Science.* 2nd edition. Englewood Cliffs: Prentice Hall.

De Klerk, Geert-Jan. 1979. "Mechanism and Vitalism. A History of the Controversy." *Acta Biotheoretica* 28 (1): 1–10. https://doi.org/10.1007/BF00054676.

Dobzhansky, Theodosius. 1964. "Biology, Molecular, and Organismic." *American Zoologist* 4 (4): 443–52. https://doi.org/10.1093/icb/4.4.443.

Falk, Raphael. 2010. "Mutagenesis as a Genetic Research Strategy." *Genetics* 185 (4): 1135–39. https://doi.org/10.1534/genetics.110.120469.

Fr. Richter, Clemens, OSA. 2015. "Remembering Johann Gregor Mendel: A Human, A Catholic Priest, An Augustinian Monk, and Abbot." *Molecular Genetics & Genomic Medicine* 3 (6): 483–85. https://doi.org/10.1002/mgg3.186.

Gerontas, Apostolos. 2013. "Reforming Separation: Chromatography from Liquid to Gas to High Performance Liquid." PhD diss., Norwegian University of Science and Technology, Trondheim.

Gerontas, Apostolos. 2020. "Chromatography." In *Between Making and Knowing: Tools in the History of Materials Research*, edited by Joseph D. Martin and Cyrus C. M. Mody, 315–26. Singapore and Hackensack, NJ: World Scientific. https://doi.org/10.1142/9789811207631_0028.

Golden, Jon C., Vinh Ho, and Vassiliy Lubchenko. 2017. "The Chemical Bond as an Emergent Phenomenon." *The Journal of Chemical Physics* 146 (17): 174982502. https://doi.org/10.1063/1.4 707.

Hempel, Carl G. 1970. "On the 'Standard Conception' of Scientific Theories." In *Analyses of Theories and Methods of Physics and Psychology, Volume 4*, edited by Michael Radner and Stephen Winokur, 142–63. Minneapolis, MN: University of Minnesota Press.

Hendry, Robin Findlay. 2010. "Ontological Reduction and Molecular Structure." *Studies in History and Philosophy of Science Part B: Studies in History and Philosophy of Modern Physics* 41 (2): 183–91. https://doi.org/10.1016/j.shpsb.2010.03.005.

Hendry, Robin Findlay. 2017. "Prospects for Strong Emergence in Chemistry." In *Philosophical and Scientific Perspectives on Downward Causation*, edited by Michele Paolini Paoletti and Francesco Orilia, 146–63. New York: Routledge. https://doi.org/10.4324/9781315638577-9.

Hettema, Hinne. 2012. *Reducing Chemistry to Physics: Limits, Models, Consequences.* Createspace.

Hettema, Hinne. 2017. *The Union of Chemistry and Physics: Linkages, Reduction, Theory Nets and Ontology.* New York: Springer. https://doi.org/10.1007/978-3-319-60910-2.

Johannsen, Wilhelm. 1911. "The Genotype Conception of Heredity." *American Naturalist* 45 (531): 329–63. https://doi.org/10.1086/279202.

Judson, Horace F. 1980. "Reflections on the Historiography of Molecular Biology." *Minerva* 18: 369–421. https://doi.org/10.1007/BF01096950.

Judson, Horace F. 1996. *The Eighth Day of Creation: The Makers of the Revolution in Biology, Expanded Edition.* Cold Spring Harbor, NY: Cold Spring Harbor Laboratory Press.

Kellenberger, Eduard. 2004. "The Evolution of Molecular Biology." *EMBO Reports* 5 (6): 546–49. https://doi.org/10.1038/sj.embor.7400180.

Kelly, Jack, Carlo Berzuini, Bernard Keavney, Maciej Tomaszewski, and Hui Guo. 2022. "A Review of Causal Discovery Methods for Molecular Network Analysis." *Molecular Genetics & Genomic Medicine* 10 (10): e2055. https://doi.org/10.1002/mgg3.2055.

Klein, Jan, and Norman Klein. 2013. *Solitude of a Humble Genius—Gregor Johann Mendel. Volume 1: Formative Years,* 91–103. Berlin: Springer. https://doi.org/10.1007/978-3-642-35254-6.

Morange, Michel. 1998. *La Part des gènes.* Paris: Odile Jacob.

Morange, Michel. 2020. *The Black Box of Biology: A History of the Molecular Revolution,* translated by Matthew Cobb. Cambridge, MA: Harvard University Press. https://doi.org/10.4159/9780674245280.

Morris, Peter J. T. 2002. *From Classical to Modern Chemistry: The Instrumental Revolution.* London: Royal Society of Chemistry.

Muller, Hermann Joseph. 1922. "Variation Due to Change in the Individual Gene." *The American Naturalist* 56 (642): 32–50. https://doi.org/10.1086/279846.

Muller, Hermann Joseph. 1926. "The Gene as the Basis of Life." *Proceedings of the International Congress of Plant Sciences* 1: 897–921.

Müller-Wille, Staffan, and Hans-Jörg Rheinberger. 2012. *A Cultural History of Heredity.* Chicago, IL: The University of Chicago Press. https://doi.org/10.7208/chicago/9780226545721.001.0001.

Nagel, Ernest. 1961. *The Structure of Science: Problems in the Logic of Scientific Explanation.* New York: Harcourt, Brace & World.

Näpinen, Leo. 2015. "The Premises and Limits of Science: The Ideas of Rein Vihalemm." *Acta Baltica Historiae et Philosophiae Scientiarum* 3 (2): 108–14. https://doi.org/10.11590/abhps.2015.2.06.

Porter, Theodore M. 2011. "Medical Science." In *The Cambridge History of Medicine,* 144–48. Cambridge: Cambridge University Press.

Porter, Theodore M. 2018. *Genetics in the Madhouse: The Unknown History of Human Heredity.* Princeton, NJ: Princeton University Press. https://doi.org/10.23943/9781400890507.

Portin, Petter, and Adam Wilkins. 2017. "The Evolving Definition of the Term 'Gene.'" *Genetics* 205 (4): 1353–64. https://doi.org/10.1534/genetics.116.196956.

Reinhardt, Carsten. 2006. *Shifting and Rearranging: Physical Methods and the Transformation of Modern Chemistry.* Sagamore Beach, MA: Science History Publications.

Rheinberger, Hans-Jörg. 1997. *Toward a History of Epistemic Things. Synthesizing Proteins in the Test Tube*. Stanford, CA: Stanford University Press.

Rheinberger, Hans-Jörg. 2012. "Internationalism and the History of Molecular Biology." In *The Globalization of Knowledge in History*, edited by Jürgen Renn, 739–44. Berlin: Max-Planck-Gesellschaft zur Förderung der Wissenschaften.

Scerri, Eric. 2012. "What Is an Element? What Is the Periodic Table? And What Does Quantum Mechanics Contribute to the Question?" *Foundations of Chemistry* 14: 69–81. https://doi.org/10.1007/s10698-011-9124-y.

Seifert, Vanessa A. 2022. "Open Questions on Emergence in Chemistry." *Communications Chemistry* 5: 49. https://doi.org/10.1038/s42004-022-00667-7.

Simon, Jonathan. 1998. "The Chemical Revolution and Pharmacy: A Disciplinary Perspective." *Ambix* 45 (1): 1–13. https://doi.org/10.1179/amb.1998.45.1.1.

Suárez-Díaz, Edna. 2016. "Molecular Evolution in Historical Perspective." *Journal of Molecular Evolution* 83 (5–6): 204–13. https://doi.org/10.1007/s00239-016-9772-6.

Vihalemm, Rein. 2007. "Philosophy of Chemistry and the Image of Science." *Foundations of Science* 12 (3): 223–34. https://doi.org/10.1007/s10699-006-9105-0.

Vihalemm, Rein. 2019. "A Story of a Science: On the Evolution of Chemistry." Special issue, *Acta Baltica Historiae et Philosophiae Scientiarum* 7 (2). https://doi.org/10.11590/abhps.2019.2.01.

Venter, Craig, and Daniel Cohen. 2004. "The Century of Biology." *New Perspectives Quarterly* 21 (4): 73–77. https://doi.org/10.1111/j.1540-5842.2004.00701.x.

Conclusion

Ave Mets

Practical realism in the philosophy of science has multifarious roots, implications, and possible applications. This volume, dedicated to Rein Vihalemm (1938–2015), the creator of this conception of science, and his legacy, expands on some of these facets.

The first part explores thinkers and grounds aligned with his thought in significant aspects that Vihalemm himself might not have realized or recognized. This exploration illustrates how philosophical views do not exist in isolation or are self-sustaining, but instead can be seen as forming networks, more or less consciously inheriting from predecessors or each other. The second part presents dissenting voices, primarily concerning metaphysical interpretations, from otherwise like-minded scholars, Rein Vihalemm's peers. The third part focuses on applications of practical realist and φ-science frameworks, including critical assessments. While most applications are in Vihalemm's own primary domains—the natural sciences—a relatively novel addition here is the application of the practical realist approach to matters pertaining to the social sciences and the humanities (psychology, religious studies), areas which do not directly stem from Vihalemm's accounts. This field certainly warrants further exploration and expansion into analogous disciplines.

The authors of the chapters often offer more articulated and detailed applications, or illustrations, of Vihalemm's frameworks than he himself provided. His more detailed studies of sciences (specifically chemistry) predate the formulation of these frameworks (though, undoubtedly, contributing to the evolution of these frameworks). Thus, this kind of specifying research, both in natural and social disciplines, would be much appreciated.

Further research should critically analyze these frameworks against specific scientific fields. Chemistry itself, Vihalemm's pet discipline and the foundation of his philosophical frameworks, appears more intricate than the frameworks can capture. Future studies in even more complex fields may necessitate refining and qualifying the practical realist approach. This also applies to the already mentioned social sciences and humanities, the specificities of the practices of which, compared to natural sciences, and also to exact sciences certainly merit greater attention. Metaphysical considerations may prove much more prominent than Vihalemm himself was prepared to acknowledge. Interestingly, his earlier notion of a world picture—a comprehension of the world, or its contents, of its basic ontology, influenced by sciences and disseminated in societies through education as a foundation of culture—indicates this prominence, alongside the term "realism." This world picture in different disciplines certainly deserves more philosophical attention.

The volume could not cover the ethical and practical implications of practical realism, which are by no means less important (rather the opposite). The abovementioned issue of world picture already clearly hints at this: it determines how we think of the natural, social, and technical-technological world around us, of ourselves and our role in this world. While practical realism clearly emphasizes the centrality of human actions and activities, it may also excessively focus on that, and, along with the assertion that the world for us *is* only via practices, it may imply a certain passivity and neglect of the world that we do not yet know to the peril of us and others (nonhuman beings). These implications need to be identified and highlighted as well.

Index

255

About the Contributors

Hasok Chang is the Hans Rausing Professor of History and Philosophy of Science at the University of Cambridge. He received his degrees from Caltech and Stanford and has taught at University College London. He is the author of *Is Water H₂O? Evidence, Realism and Pluralism* (2012), *Inventing Temperature: Measurement and Scientific Progress* (2004), and *Realism for Realistic People: A New Pragmatist Philosophy of Science* (2022). He is a cofounder of the Society for Philosophy of Science in Practice and the Committee for Integrated History and Philosophy of Science.

Jaana Eigi-Watkin is a research fellow in the philosophy of science at the Institute of Philosophy and Semiotics, University of Tartu, Estonia. She received her PhD degree in philosophy from the University of Tartu in 2016. She is interested in the social aspects of science, as they are widely understood. Independently, she has published on the social aspects of scientific objectivity, the epistemic and political facets of the democratization of science, and the role of nonscientists in science. As a member of the research group led by Dr. Endla Lõhkivi, she has coauthored papers that used interviews with researchers to qualitatively analyze the epistemic implications of work cultures in various disciplines.

Apostolos Gerontas, PhD, studied food chemistry and microbiology at the Aristotle University of Thessaloniki in Greece; history of science at the University of Stuttgart, Germany; and history of science and science education at the Norwegian University of Science and Technology in Trondheim. He has been twice a fellow of the Science History Institute in Philadelphia, PA, and a visiting researcher at the American College of Greece in Athens. He teaches at the Coburg University of Applied Sciences, Germany, and the Norwegian University of Science and Technology in Trondheim, Norway.

David Hommen received his doctorate and habilitation in philosophy from Heinrich Heine University Düsseldorf, Germany. He currently works there as a privatdozent and leads a research project on the "Presuppositions of Frame Theory in the History of Philosophy." He has published on a wide variety of topics, including the mind-body problem; the causality of omissions; the structure of conceptual representations; the metaphysics of objects, properties, and kinds; and the philosophy of Ludwig Wittgenstein. His current research focuses on the relationship between language, thought, and reality, and on the possibility and nature of knowledge in philosophy and science. He is the review editor for the journal *History of Philosophy & Logical Analysis*.

Juho Lindholm has an MSc degree in technology and an MA degree in philosophy from Finland. He is currently a PhD student at the University of Tartu. He studies pragmatism, naturalism, philosophy of technics, and postphenomenology.

Olimpia Lombardi is an electronic engineer and has a PhD in philosophy from the University of Buenos Aires. She is a senior researcher at CONICET (National Scientific and Technical Research Council, Argentina), head of the Group of Philosophy of Sciences at the Institute of Philosophy, University of Buenos Aires, member of the *Academie Internationale de Philosophie des Sciences* and of the Foundational Questions Institute. Lombardi is also a research associate at the Centre for Philosophy of Natural and Social Science of the London School of Economics and Political Science (LSE), Charter Honorary Fellow of the John Bell Institute for the Foundations of Physics, member of the Executive Committee of the International Society for the Philosophy of Chemistry (ISPC), and associate editor of *Foundations of Physics*, *Foundations of Chemistry*, *Philosophy of Physics*, and *Hyle—International Journal for Philosophy of Chemistry*. Lombardi's areas of interest are foundations of statistical mechanics, the problem of the arrow of time, interpretation of quantum mechanics, the nature of information, and philosophy of chemistry.

Endla Lõhkivi is an associate professor of philosophy of science at the University of Tartu, Estonia. She studied chemistry in Tartu, thereafter philosophy in Göteborg, Sweden, and Tartu. Her research interests include history and philosophy of science, especially chemistry, science as culture, interdisciplinarity and conceptual change, and expanding and building upon the practice-based approach started by Vihalemm. She has participated in international networks, led projects, taught extensively in Tartu and abroad, and was the chair of Philosophy of Science in Tartu for a long time.

Ave Mets is a research fellow of philosophy of science at the Institute of Philosophy and Semiotics, University of Tartu, Estonia. She completed all her academic degrees in Tartu under Rein Vihalemm's supervision and was a junior research fellow at the Human Technology Center of the RWTH Aachen University, a postdoctoral fellow at Belarus State University, and a Fulbright visiting scholar at Washington State University. She has taught philosophy of science and technology at many universities internationally (Reykjavik, Oulu, Istanbul, Kaunas, etc.) and published on various topics (practice-based philosophies of chemistry, measurement, world picture) in both academic and popular venues.

Bruno Mölder studied philosophy at the University of Tartu (under Rein Vihalemm's supervision), the University of Cambridge, and the University of Konstanz. He has been a visiting fellow internationally (in Aarhus, Denmark and Oxford, UK) and received several awards for his academic work. His academic interest lies in the philosophy of mind, especially the interpretivist stance, time consciousness, and social cognition. He is currently a professor of the philosophy of mind and the head of the Institute of Philosophy and Semiotics at the University of Tartu.

Peeter Müürsepp is an associate professor at Tallinn University of Technology, Estonia, and received his PhD in philosophy from Vilnius University, Lithuania. He has been a visiting research fellow at the LSE, UK; New School for Social Research (New York City); Roskilde University Centre, Denmark; and the Helsinki Collegium for Advanced Studies, Finland. Dr. Müürsepp is the editor-in-chief of two SCOPUS indexed journals, *Acta Baltica Historiae et Philosophiae Scientiarum* and *ICON*. He is an editorial board member of various international and domestic publications. Dr. Müürsepp has been the chairperson of the Estonian Association for the History and Philosophy of Science since 2011 and is currently also the president of the Baltic Association for the History and Philosophy of Science. He is a corresponding member of the International Academy of the History of Science.

Alexander Pechenkin graduated from Mendeleev Institute of Chemistry and Technology (1963) and the Department of Applied Mathematics at Lomonosov Moscow State University (1975). He received his PhD from the Institute of Philosophy of the Academy of Sciences (1968). Alexander Pechenkin has had internships at the following scientific establishments:

Harvard University, USA; Cambridge University, UK; the Central European University in Budapest, Hungary; the Chemical Heritage Foundation, USA; *Forschungsinstitut am Deutschen Museum* in Munich, Germany; and Institute for Advanced Studies in Helsinki, Finland. Alexander Pechenkin is a professor at Lomonosov Moscow State University's School of Philosophy and a senior researcher for S. Vavilov Institute for the History of Science and Technology, Russia.

Sami Pihlström is a professor of philosophy of religion at the University of Helsinki, Finland. He is also, among other things, the former president of the Philosophical Society of Finland. He has published widely (more than twenty books and hundreds of articles) on, for example, pragmatism, the problem of realism, ethics, metaphysics, and the philosophy of religion. His recent monographs include *Why Solipsism Matters* (2020), *Pragmatist Truth in the Post-Truth Age: Sincerity, Normativity, and Humanism* (2021), and *Toward a Pragmatist Philosophy of the Humanities* (2022).

Klaus Ruthenberg studied chemistry and philosophy in his hometown Berlin and Göttingen with doctorates in both disciplines. After some years in research institutes, universities, and the chemical industry, he became a professor of chemistry at the Coburg University of Applied Sciences in Germany in 1989. Like Rein Vihalemm, he is a cofounder of the ISPC and has held visiting positions at several universities abroad, such as Bradford, UK; Leuven, Belgium; and Cambridge, UK. His main research fields in the history and philosophy of chemistry consist of chemical kinds, chemical concepts, and philosophizing chemists. His latest extensive publication is the monograph *Chemiephilosophie* (2022).

Kenneth R. Westphal is a lifetime member of the Academia Europaea (https://ae-info.org/). Recently, he retired to Trieste, Italy. His research focuses on the character and scope of rational justification in nonformal, substantive domains, both moral and theoretical. His extensive publications include *Kant's Transcendental Proof of Realism* (2004), *Grounds of Pragmatic Realism: Hegel's Internal Critique and Transformation of Kant's Critical Philosophy* (2018), *Kant's Critical Epistemology: Why Epistemology must Consider Judgment First* (2020), and *Kant's Transcendental Deduction of the Categories: Critical Re-examination, Elucidation & Corroboration* (2021).